MMS

Technologies, Usage and Business Models

MMS

Technologies, Usage and Business Models

Daniel Ralph
BTexact Technologies, Ipswich, UK

Paul Graham
eServGlobal, Ipswich, UK

John Wiley & Sons, Ltd

Other Wiley Editorial Offices

John Wiley & Sons Inc., 111 River Street, Hoboken, NJ 07030, USA

Jossey-Bass, 989 Market Street, San Francisco, CA 94103-1741, USA

Wiley-VCH Verlag GmbH, Boschstr. 12, D-69469 Weinheim, Germany

John Wiley & Sons Australia Ltd, 33 Park Road, Milton, Queensland 4064, Australia

John Wiley & Sons (Asia) Pte Ltd, 2 Clementi Loop #02-01, Jin Xing Distripark, Singapore 129809

John Wiley & Sons Canada Ltd, 22 Worcester Road, Etobicoke, Ontario, Canada M9W 1L1

Wiley also publishes its books in a variety of electronic formats. Some content that appears in print
may not be available in electronic books.

British Library Cataloguing in Publication Data

A catalogue record for this book is available from the British Library

ISBN 0-470-86116-9

Project management by Originator, Gt Yarmouth, Norfolk (typeset in 11/13pt Palatino)
Printed and bound in Great Britain by Antony Rowe, Chippenham, Wiltshire
This book is printed on acid-free paper responsibly manufactured from sustainable forestry
in which at least two trees are planted for each one used for paper production.

Contents

Multimedia Messaging Service   Daniel Ralph and Paul Graham
© 2004 John Wiley & Sons, Ltd   ISBN: 0-470-86116-9

About the Authors

Daniel Ralph is an engineering manager at BTexact Technologies, where he is responsible for a number of projects in the mobile applications arena. He currently consults and project-manages application development, application integration and systems migration, and is also interested in the wider commercial, social and political implications associated with the impact of the knowledge society.

Daniel received his masters degree in Telecommunications Engineering at University College London. He also holds a BSc(Hons) in Computer Science from the Open University. He is a member of the British Computer Society and is a chartered engineer. He has authored a number of journal papers and presented at conferences on the subject of delivering services via mobility portals and technologies of the mobile Internet.

Paul Graham is an engineering manager at eServGlobal, where he is responsible for a number of projects associated with the development of Intelligent Network Services and mobile applications.

Paul received his master's degree in Telecommunications Technology at Aston University. He also holds a BEng in Electronics from Southampton University. He is a member of the IEE and has vast experience in the telecommunications industry, having served three years at BT's research facility at Martlesham Heath (near Ipswich), followed by five years working for G8 Labs/eServGlobal.

Multimedia Messaging Service   Daniel Ralph and Paul Graham
© 2004 John Wiley & Sons, Ltd   ISBN: 0-470-86116-9

This includes periods working for Stratus, Ascend and Lucent Technologies in the USA and throughout Europe. His current interests include next-generation Internet services and mobile services. He specializes in billing systems and data services.

How This Book Is Organized

We have tried to organize the chapters in this book in a modular fashion to accommodate the interests and backgrounds of as many readers as possible. We have also tried to keep a fairly informal style, while doing our best to introduce every single technical term and acronym (after having read through the first chapter, you will realize that there are quite a few). There is also a glossary of terms provided at the end of the book to save you from having to remember where each term was first introduced – especially if you do not plan to read every chapter. In general, we have assumed a fairly minimal technological understanding.

If you are a manager who reads technology articles in publications such as *Computing* or *The Financial Times*, you should have no problem following and will hopefully enjoy the many service-oriented discussions we have included. On the other hand, if you are a developer who is already familiar with mobile communication technologies, Internet protocols or encryption, you might want to just skim through the first few pages of chapters where we briefly review some of the basics associated with these technologies. In general, we have tried to stay away from deep technological discussions – the Third Generation Partnership Project (3GPP) standards and Internet Engineering Task Force (IETF)

Multimedia Messaging Service   Daniel Ralph and Paul Graham
© 2004 John Wiley & Sons, Ltd   ISBN: 0-470-86116-9

RFC (Request For Comments) provide sufficient technical detail on each of the technologies we cover, some of which are listed in the references at the end of this text. Instead, our goal is to give you a broad overview of the many business models, technologies and services of Multimedia Messaging Services (MMSs). In the process we try to show how technologies and standards impact the business models of players across the value chain, and how existing and new business models and usage scenarios also drive the development of new technologies and standards.

Specifically, the book is organized in three parts.

PART I MOBILE MESSAGING BUSINESS CHALLENGES

Chapter 1 Multimedia Messaging Overview

Here, we attempt to understand the MMS proposition. We take a look at some of the services and business models of different players across the value chain. In the process, we review the drivers behind MMS and explain how they differ from the Short Message Service (SMS) and the Enhanced Message Service (EMS). Hopefully, by the time you reach the end of this chapter, you will have realized that multimedia messaging is very different – not just in technology terms but also in terms of services, usage scenarios, players and business models.

Chapter 2 The Multimedia Messaging Value Chain

In Chapter 2 we take a look at the many different categories of players found across the multimedia messaging value chain. We try to understand the context within which they operate, and how the threats and opportunities of multimedia messaging impact their business models. This includes a discussion of mobile operators, content providers, handset manufacturers and equipment providers.

PART II   THE TECHNOLOGIES OF MULTIMEDIA MESSAGING

Chapter 3   A Standards-based Approach

Chapter 3 starts with an introduction to the major standards activities that underpin the implementation of MMS and follows on with an overview of the main MMS and WAP (Wireless Application Protocol) standards. Rather than getting into the detailed specifics of each release of these standards, our discussion focuses on major features, such as video streaming, billing, content provision, interoperability, message delivery and storage. We also look at the roadmap associated with the future developments envisaged in *MMS Release 6*.

Chapter 4   Application Layer

In this chaper we begin by discussing the variety of terminal types supporting the MMS standard. This leads to the issues of content adaptation, device support requirements, presentation and lastly provisioning of services. An introduction to the Synchronized Multimedia Integration Language (often referred to as SMIL) enables an understanding of the capability of MMS that extends it far beyond the sending of static images. Then follows guidance on how to develop content for deployment on MMS devices. Content providers will be required to use a specific interface provided from the mobile operator's network; this will support billing, authentication and audit as a minimum and is an important area for the deployment of MMS. The implications for WAP gateway infrastructure are discussed, and the chapter concludes with an overview of MMS streaming, an enhancement to support video playback to MMS devices.

Chapter 5   Network Layer

In this chapter we detail the enabling network technologies behind MMS. It commences with an outline of each component, from the MMS server/relay, the database interactions, network interoperability and operation with billing and accounting systems. The chapter covers billing systems and the methods in which MMS will rate and charge for usage, as it is felt this is a significant and important topic for operators and service providers. Having an understanding of the standards available will allow operators to develop novel methods in which to charge for the usage of the service and give a competitive edge. Example billing scenarios are covered, as well as ideas for realizing them. It also examines new concepts such as the MMBox, a standards-based method to enable persistent network storage of multimedia messages, and concepts such as reply charging, which provides a method to enable senders of multimedia messages to pay for replies to them, very important in the world of advertising and marketing. The chapter includes a discussion about developing the required software components, covering the concept of using open source software and its viability within the MMS environment. The chapter concludes with the topic of interoperability, important for message transfer between different networks and roaming.

PART III   MULTIMEDIA MESSAGING SERVICES TODAY AND TOMORROW

Chapter 6   Multimedia Messaging Services Today

Part III is where all the pieces come together. We begin by looking at MMSs available today. Our discussion covers both consumer services and business applications. Here, we review mobile portals as well as a number of specific services such as location-based services, retailing and mobile entertainment. In each case we look at emerging usage scenarios and business opportunities and compare the approaches taken by different players.

We conclude the book with future recommendations, tackling such issues as interoperability and evolution.

Chapter 7   Future Recommendations

Many of today's MMSs remain very limited when it comes to offering customers services that are directly relevant to their preferences, location or other context-based attributes (e.g., weather, travel plans, people you are with and so forth). In this chapter we introduce new solutions aimed at facilitating personalization and context awareness to mobile services. This includes a discussion of the 3GPP Personal Services Environment (PSE) and Generic User Profile (GUP), which aims to provide a single repository for all customers' personal preferences. Similar industry efforts have been launched, including Microsoft Passport and Sun Liberty Alliance.

FURTHER READINGS

For those of you interested in further exploring the topics discussed in this text, we have included a list of references and websites at the end of the book.

Finally, a summary of each chapter is provided along with a guide for how to read this book depending on your specific background and interests. As you will see, we have tried to keep the text modular to accommodate as broad a range of readers as possible. We hope you find this easy to navigate. Enjoy the book!

Acknowledgements

One of the primary aims of this book is to share some of the learning experiences gained over the last few years from developing and integrating technologies required in the deployment of MMSs for mobile operators. Most of my experience in the arena of the mobile Internet has been gained at the application layer, developing and integrating applications and infrastructure for mobile operators, large corporates and banks. With my colleagues I have worked on projects including: WAP banking, Manx 3G Showcase, O2 Unified Messaging and many bespoke business applications.

I would like, in particular, to thank Andy Pearson (BTexact) for his support and for giving me the time to work on this book while managing my engineering responsibilities and Ian Dufour (BTexact), without whose encouragement and support this project would not have been possible. I would like to thank my co-author, Paul Graham (eServGlobal) for his invaluable input and experience in the areas of billing and networks. This has enabled a rounded perspective of the topic. The editorial team at Wiley have been fantastic. Special thanks to Sally Mortimore, my editor, for giving me the opportunity to have my work published. Thanks also to the other members of the team, including Birgit Gruber and the reviewers. I am indebted to Ian Harris (Chairman 3GPP/TSG T2) and Dario Betti (Ovum) for discussions on the opportunities and

Multimedia Messaging Service   Daniel Ralph and Paul Graham
© 2004 John Wiley & Sons, Ltd   ISBN: 0-470-86116-9

challenges presented by MMS and to my BT exact colleagues Julie Harmer, Xin Guo and Stephen Searby for their engineering support. Thanks also to Gwenael Le Bodic for his detailed technical analysis of the topic of MMS and the 3GPP specifications.

Last but not least, I would like to thank my wife Sarah for putting up with my disappearing acts to complete another chapter. This book gives me the opportunity to thank her for her love and unfailing support.

Daniel Ralph
Ipswich, UK
daniel.ralph@bt.com

I would like to thank all those who have provided support over the years, including Mike Scott, Adrian Seal and Andrew Taylor. To my co-author Daniel, for providing the foundation for the book and for allowing us to continue our professional relationship, which has continued to be successful. I've worked on many enoyable projects with KPN, Mobistar Belgium and BT, which have provided invaluable experience. Special thanks to Robert Edwards for offering continued superior technical advice and for reviewing the work. And finally thanks to Celine Hamel for her complete support and understanding during the project and throughout my career.

Paul Graham
eServGlobal
paul.graham@eservglobal.com

Part I

Mobile Messaging Business Challenges

1

Multimedia Messaging Overview

1.1 INTRODUCTION

Multimedia Messaging Service (MMS) is the evolution of basic text messaging services into a wide range of multimedia content and services delivered to a mobile device. It involves new and existing technologies, promises innovative services and is supported by proven business models. It is quite different from traditional Internet services as its integration with billing systems allows for per message billing and premium content services, all delivered without the need for additional subscriptions.

With the improved capability of mobile devices, in terms of increased processing power and bandwidth, mobile operators are making the transition from supporting voice-only communication to embracing data-driven applications (or data services). The most notable of these data services is the Short Message Service (SMS); during 2002 globally in excess of 24 billion messages were sent in one month. This trend is set to continue with worldwide estimates of 360 billion SMS messages to be sent in 2003 (**http://www.gsmworld.com/news/press_2002/press_10_pl47.shtml**). This phenomenon has driven UK mobile operators' data revenue up by over 10%. It is this strong growth (faced with declining voice revenues) that gave some operators in the European mobile industry unrealistic expectations of consumer and corporate spend, often

Multimedia Messaging Service   Daniel Ralph and Paul Graham
© 2004 John Wiley & Sons, Ltd   ISBN: 0-470-86116-9

referred to as Average Revenue Per User (ARPU). The expectation of higher revenues from data services in turn led to the $116 billion spent on 3G (Third Generation) licenses across Europe by mobile operators.

In Japan the NTT DoCoMo i-mode service was tremendously successful in engaging the consumer market and provided information, location, entertainment and gaming services to mobile devices with colour screens and 'always-on' data connections. This led to over 37 million subscribers within three years and more recently attempts to recreate this business model outside Japan (**http://www.mobilemediajapan.com**/). The Japanese operator J-Phone launched a photo-messaging service in 2000 and, while the image quality is of low resolution, they achieved over 7 million handset subscribers by 2002 (**www.vodafone.com** media centre).

In writing this book, our objective is to highlight the impact of MMS on the infrastructure and business models of mobile operators, service providers and content providers. The effect on business models is directly related to the usability and ease of configuration of the service, not forgetting the pricing and billing strategy adopted by the service provider, content provider and mobile operator. We follow this up with an overview of MMS technology components: application presentation to mobile terminals, interface definitions and network elements essential in developing and delivering MMS. Furthermore, the mobile operator infrastructure requires fundamental changes to mail and billing systems and the complexity of integrating high transaction services should not be underestimated.

In this chapter we describe the fundamentals of the service offering behind MMS and examine the challenges facing the mobile industry in the growth of data services. This is supplemented by a detailed assessment of the challenges facing MMS deployments and associated revenue generation.

1.2   WHAT IS MULTIMEDIA MESSAGING?

SMS, Enhanced Message Service (EMS) and MMS are non-proprietary, open standards for Person-to-Person (P2P) and Machine-to-Person (P2M) messaging over a signalling channel of a GSM (Global System for Mobile) network (in the case of SMS) or a packet-based mobile network (in the case of EMS and MMS). Unlike proprietary technologies, such as Nokia Smart Messaging or the Cybiko, they have been developed and maintained within recognized industry forums.

When comparing functionality an evolutionary path can be seen by means of three standards:

- SMS enables mobile phone users to send short, plain-text messages to other mobile phone users;

- EMS enables mobile phone users to send longer text messages, plus simple graphics and sounds, to other mobile phone users;

- MMS enables mobile phone users to send formatted text messages (theoretically of any length), plus graphics, photographic images, and audio and video content, to other mobile phone users and to email users.

The EMS and MMS are specifications being developed by the 3G Partnership Project (3GPP) and the technical realizations of MMS are prepared by the Open Mobile Alliance (previously known as the WAP Forum). There may be two separate stages of evolution from SMS, with users being migrated first from SMS to EMS, and then from EMS to MMS (Svensson, 2001). The principal features are summarized in Figure 1.1.

Because each standard provides clear additional features to its predecessor, mobile messaging can follow a simple transition from SMS, to EMS and finally to MMS. However, even though MMS is based on a very different set of technologies from the other two standards, the migration path between them will use many of the existing systems (e.g., SMSC [SMS Centre], WAP Gateway, Billing Systems and Home Location Register [HLR]).

There are also factors relating to the timing of the standards and to industry politics, which may limit the significance of EMS in the overall development of mobile messaging. Significantly, Nokia have not committed to implement the EMS standard on its range of phones and this will severely impact the long-term availability of the service. So, while some migration through EMS will occur, as shown on p. 7, in the main the migration that will take place in mobile messaging over the next few years will likely be straight from SMS to MMS.

With MMS, subscribers are able to compose and receive messages ranging from simple plain text messages, as found in SMS

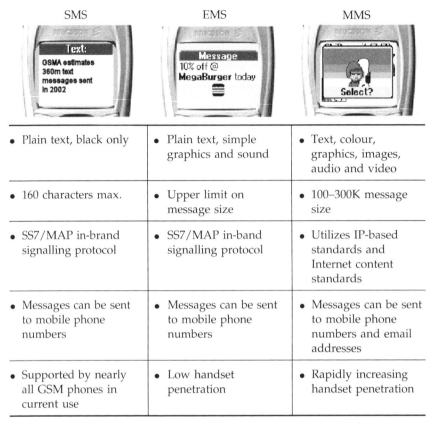

SMS	EMS	MMS
• Plain text, black only	• Plain text, simple graphics and sound	• Text, colour, graphics, images, audio and video
• 160 characters max.	• Upper limit on message size	• 100–300K message size
• SS7/MAP in-brand signalling protocol	• SS7/MAP in-band signalling protocol	• Utilizes IP-based standards and Internet content standards
• Messages can be sent to mobile phone numbers	• Messages can be sent to mobile phone numbers	• Messages can be sent to mobile phone numbers and email addresses
• Supported by nearly all GSM phones in current use	• Low handset penetration	• Rapidly increasing handset penetration

SMS = Short Messaging Service; MMS = Multimedia Message Service; SS7/MAP = Signalling System 7/Mobile Application Part; GSM = Global System for Mobile communication; IP = Internet Protocol.

Figure 1.1 SMS to MMS evolution
Source: Vodafone website

and EMS, to complex multimedia messages (similar to Shockwave Flash animations found on websites). A Multimedia Message (MM) can be structured as a slideshow, similar to a Microsoft PowerPoint presentation. Each slide is composed of elements such as text, audio, video or images and the period of time each element is displayed or played (within the slide) can be determined by the content developer. Support is available to send messages to distribution lists and manage the delivery and read–reply reports. Messages can have different class types, such as advert, information or personal. The concept of message notification allows immediate or deferred retrieval of messages.

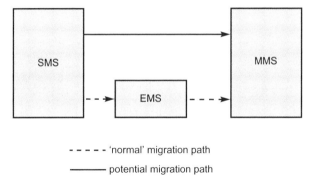

- - - - - 'normal' migration path

——————— potential migration path

Figure 1.2   SMS to MMS migration path

SMS = Short Message Service; EMS = Enhanced Message Service; MMS = Multimedia Messaging Service

In order to cope with the storage capacity limitations of mobile devices, MMS utilizes the concept of a persistent network-based store, often referred to as a content album (or its technical term the MMBox).

Although MMS supports multiple content types, the devices will inherently differ in the level of support provided for each media format. The use of content adaptation (e.g., the reformatting of a graphic so it displays entirely on a mobile phone screen) is to ensure good user experience is provided to the appropriate device capability. The overall network architecture for MMS is outlined in Figure 1.3. The basis of this architecture forms the discussion throughout this book and will prove a useful reference during your reading.

1.2.1   Vodafone Live! service

In October 2002 Vodafone Live! (the platform for Vodafone Group's new portfolio of mobile data services) was launched in the UK, Germany, Italy, Spain, Portugal and The Netherlands; other countries will follow, including Ireland, Sweden, Greece, Australia and New Zealand (**www.vodafone.com** media centre). Vodafone Live! is Vodafone Group's consumer mobile data platform. It has been developed since 2001 by a number of research groups within Vodafone. Among others, it includes elements of Vizzavi's technology and work from the development labs of Vodafone. It conforms to the industry-wide M-Services guidelines

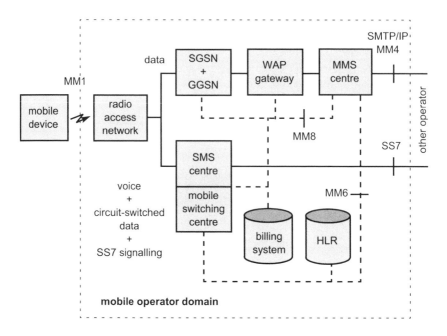

Figure 1.3   Overall network architecture
SMTP = Simple Mail Transfer Protocol; IP = Internet Protocol; SGSN = Serving GPRS Support Node;
GPRS = General Packet Radio Service; GGSN = Gateway GPRS Support Node; WAP = Wireless
Application Protocol; MMS = Multimedia Messaging Service; SMS = Short Message Service;
SS7 = Signalling System No. 7; HLR = Home Location Register; MM1 = interface to MMS relay/
server; MM4 = interface between other operators' MMS Centres; MM6 = interface to HLR;
MM8 = interface between the MMSC and the billing engine

(GSM Association, 2001). These elements have been brought to-
gether in a platform that will be deployed across the Vodafone
Group, incorporating new mobile services such as picture mes-
saging (as MMS is often referred), Java games downloads, Wireless
Application Protocol (WAP) information services and Instant
Messaging (IM).

These services will be offered through a common user interface
across all phones and across all Vodafone networks and to this end
the operator has deployed its own software onto a number of
Vodafone Live!-enabled mobile phones. As has been highlighted
by a number of other operators, any successful mobile data
service must work "out-of-the-box". Vodafone Live! is set up on
a phone from the moment it is purchased and can be configured
Over The Air (OTA) from their website **www.vodafone.co.uk/live**
(Figure 1.4).

Figure 1.4   Vodafone Live!
Source: Vodafone website

Vodafone Live! has two core elements – a new network-based technology platform and a new user interface, which will offer commonality to all Vodafone Live! phones.

Vodafone Live! presents the user with an icon-driven user interface that should go some way to achieving a common look and feel on all devices. A simple menu structure makes use of colour screens and is designed to enable quick and simple access to new services with only one or two keystrokes. The new user interface has two key elements. First, there is a phone menu – this is similar to the menu structure that many users are already comfortable with, as it has similarities to the Nokia menu system. The first icon on this menu enables access to the phone's camera. Meanwhile, the messaging part of the menu is accessed from a soft key on the start-up screen. The second part of the user interface is accessed by pressing the "Vodafone Live!" hotkey. This is a network-based WAP menu, but instead of appearing like a traditional text-only WAP menu, it is icon-driven, allowing quick access to games downloads, billing information, messaging (both network-based Instant Messaging [IM] and email) and the more traditional WAP content – Vizzavi, Vodafone's previous portal offering can still be accessed (CSFB, 2002).

The online content at Vodafone Live! is controlled by Vodafone and, while users can browse to WAP sites not directly linked from their portal, using the keypad to input external sites is not easy and is often offputting to the point that users do not bother.

However, the Vodafone Live! graphical user interface is not unique. Phones such as the Nokia 7650 and SonyEricsson T68i offer icon-driven interfaces developed by the operating system supplier, Symbian, who is trying to encourage licensing of the Symbian Series 60 and v7 smartphone user interface to other manufacturers. To date, Samsung, Siemens and Panasonic have already committed to Series 60, although at the time of writing no devices have yet been released (Ives, 2002).

The technology platform should be transparent and seamless to the Vodafone user when roaming. According to the company website, it will allow seamless access to a range of new and existing mobile data services, not only from a user's home network but across the Vodafone footprint as well. This market dominance should help assist the deployment of interoperable services across all operators. At the core of the new services is a picture messaging service. This is built on an MMS platform from Ericsson. Vodafone (UK) are offering MMS at 36p (€0.57) per message (from early 2003). Vodafone has frequently emphasized the importance of picture messaging interoperability. The three Vodafone Live! devices are interoperable from launch and offer interoperability with TIM in Italy and T-Mobile in Germany. New interoperability agreements are being press-launched through 2003.

1.2.2 Services on offer

- picture messaging – Vodafone Live! phones are all MMS-enabled, integrated camera phones;

- instant messaging – a new service for Vodafone Live! from Followap;

- email – to replace Vizzavi email;

- polyphonic ringtones – commonplace in new terminals;

- downloadable games – downloadable Java games can be added to phone memory;

- location-based services – determine the phone's location to offer "find and seek" services;

- portal services – the Vizzavi services accessed from Vodafone Live!

 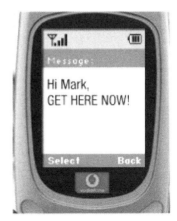

Figure 1.5 Sending an MMS using Vodafone Live!
Source: Vodafone website

In terms of revenue, the company sees Vodafone Live! as an important part of its strategy for data to reach 20% of revenues from 2004. For Vodafone Live! the issue of device interoperability is being tackled with a pragmatic approach to what is known as "legacy terminal support". An MM sent to an "old" (non-MMS-enabled) phone will be received as an SMS with a link to a website where the picture can be viewed.

Picture messaging and downloading games (written in Java, J2ME$_2$ [Java 2 Micro Edition]) are central to Vodafone Live! In the UK all operators, including Orange, mmO$_2$ and T-Mobile, have launched a picture messaging service, while in October 2002 mmO$_2$ also launched a suite of Java games that can be downloaded to the phone OTA. In Vodafone's other European markets –

Germany, Italy and Spain – competing operators have all launched both MMS and Java services. Vodafone Live! attempts to extend these services by offering them across its European footprint and with a new and integrated user interface. The ability to roam and continue to use these services will be essential in order to support the delivery of electronic postcards while visiting a holiday destination, or the downloading of games while waiting for a plane.

Vodafone is not the only company developing new user-friendly suites of mobile data services; as mentioned earlier there is an industry-wide set of guidelines that ensures these services share the same capabilities. The M-Services guidelines were announced by the GSM Association in June 2001 and updated in February 2002. They are not intended as a standard, but more a framework for a set of key technologies (e.g., MMS and Java Application Download), a common look and feel to a user interface and widespread interoperability within which operators are able to develop their own services. Vodafone Live! is really their interpretation of M-Services under the Vodafone brand. Originally, M-Services was being driven by TIM and Openwave; other vendors and operators have since become involved. For example, the Nokia Series 60 smartphone platform supports the M-Services guidelines. The GSM Association M-Services Special Interest Group Committee gives a further indication of the widespread support for the guidelines.

1.2.3 J-Phone Sha-Mail service

Vodafone-controlled Japanese operator J-Phone launched its Sha-Mail service in November 2000. The service allows users to take a photo using their mobile phone and send it to another subscriber; this was the first of its kind in Japan. To use the service, subscribers must have a handset with an integrated camera; however, they cannot send photos to non-J-Phone subscribers.

By October 2001, seven models were on the market, produced by Mitsubishi, Sanyo, Sharp and Toshiba. The latest version by Sharp opens like a clamshell and includes a Thin Film Transistor (TFT)-LCD screen display with up to 65,536 colours and a resolution of 120×160 pixels. The price for the terminal is between ¥10,000 and ¥20,000 ($75 and $150), and it is therefore comparable with other high-end terminals such as NTT DoCoMo's i-mode,

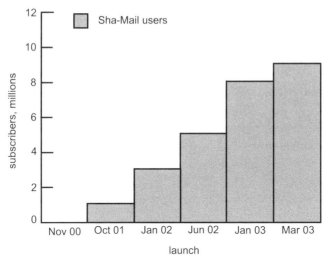

Figure 1.6   Sha-Mail growth in subscribers
Source: Company website

i-shot, i-appli and the 3G "Freedom of Multimedia Access" (FOMA) devices. By January 2002, the Japanese carrier had sold 3 million Sha-Mail-enabled handsets in an astonishing surge of growth. The bulk of this growth occurred in the second half of 2001 and propelled J-Phone from the number three to number two spot in terms of wireless data subscribers.

Sha-Mail was very well received by J-Phone's user base, which has a strong youth component. The "viral" effect of teenagers seeing their friends using the phones to take and send snapshots cannot be underestimated.

The success of Sha-Mail, highlighted in Figure 1.6, has also attracted attention from third-party service providers. For example, in February 2002 AichiTelevision Broadcasting launched an interactive TV programme called Syamekke, which allows users to participate by sending in photos and accompanying text messages on a weekly theme. The photos are shown on the programme, as well as being stored on the menu of J-Phone's J-Sky Internet service and AichiTelevision's website.

It is particularly interesting that the low picture resolution offered by the Sha-Mail service does not seem to have acted as a major inhibitor to service adoption.

J-Phone is also migrating their photo messaging service to the MMS standard; this is part of the Vodafone strategy to enable the same devices and services to operate on all their networks across their global footprint.

1.2.4 Openwave Multimedia Messaging Service Centre

The Openwave Multimedia Messaging Services Centre (MMSC) is a high-capacity, massively scalable service platform that brings the multimedia messaging capabilities of the Internet to mobile devices, while retaining the immediacy and simplicity of the mobile SMS that is popular in many parts of the world today. It enables handset users to send and receive messages that contain text, music, graphics, video and other media types – all in the same message. The Openwave MMSC is built on open standards (WAP Forum [WAPF] and Third Generation Partnership Project [3GPP] MMS) and runs on all WAP-capable mobile networks (Figure 1.7).

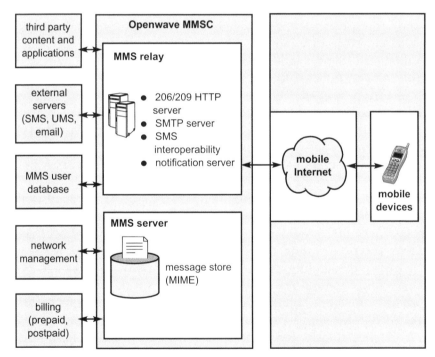

Figure 1.7   Openwave MMSC

SMS = Short Message Service; UMS = Unified Messaging System; MMS = Multimedia Messaging Service; HTTP = Hyper Text Transfer Protocol; SMTP = Simple Mail Transfer Protocol; MI-ME = Multipurpose Internet Email Extension. Source: Vodafone website

Additionally, interoperability with SMS ensures that legacy handset users are included in the MMS community and encourages mobile phone upgrades.

The distributed Internet Protocol (IP)-based architecture of Openwave MMSC utilizes low-cost, high-volume hardware components and benefits from the simplicity, efficiency and cost savings associated with operating a single message store and directory. The distributed nature of the MMSC allows operators to cost-effectively scale deployment as demand increases – capacity expansion need only be applied to those components impacted by the anticipated increase in subscribers or usage patterns. Large mobile client email and photo messaging deployments in Japan have leveraged Openwave technology and standards, such as WAP, Simple Mail Transfer Protocol (SMTP), Multipurpose Internet Email Extension (MIME), Lightweight Directory Access Protocol (LDAP) and Hyper Text Transfer Protocol (HTTP) access.

1.2.5 MMSC for carriers

The Openwave MMSC is fully IP-based and is compliant with 3GPP and WAP/OMA (Open Mobile Alliance) MMS. Also, it is designed with a commitment to interoperate with third-party technologies and systems over open interfaces. It is designed for the demanding operator market, the MMSC supports millions of subscribers with fast, reliable performance over a variety of network interfaces. The Openwave MMSC:

- is 3GPP and WAP Forum MMS-compliant;
- is built on proven Openwave technologies;
- supports LDAP, SMTP, MIME and WAP industry standards;
- is carrier class: scalable and reliable;
- provides high availability and high performance;
- is delivered as a total solution with prepaid billing, postpaid billing, per-message revenue, autoprovisioning and an operations and administration infrastructure;
- is supported by an ecosystem of systems integrators, billing technology vendors, and MMS content and application value-added service providers.

Designed specifically for communication service providers, the Openwave MMSC is enabled on a messaging platform that is flexible enough to adapt to evolving standards and mobile operator needs. It is also reusable for other classes of messaging, including unified messaging, instant messaging and email. The Openwave MMSC leverages the strengths of the Openwave mobile access gateway (an implementation of a WAP gateway) and interoperates with third-party WAP gateways as well.

1.2.6   MMSC for end-users

The Openwave MMSC allows end-users to enjoy MMS messaging applications including text, photo images, music and video clips over their wireless devices. The MMSC also provides an external application interface (known as the MM7 interface) for introducing a variety of add-on applications to make user experience even more compelling. This interface is discussed in more detail in Chapter 4 and the applications it supports are considered in Chapter 6.

1.3   GROWING MOBILE DATA REVENUES

Now that we have looked at the initial deployments of MMS, it is time to gain a better understanding of the drivers behind it. Broadly, they fall into four categories:

- device availability;

- convergence;

- towards 3G;

- supporting technologies.

1.3.1   Device availability

The tremendous growth in mobile subscribers from 1997 to 2002 has started to plateau at 1.3 billion users worldwide. As penetration has reached 70% in many major markets around the world, this is generally thought to be the saturation point (Figures 1.8 and 1.9).

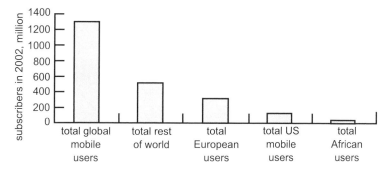

Figure 1.8 Number of world subscribers (millions) by region at December 2002

Source: EMC, **http://www.emc-database.com**

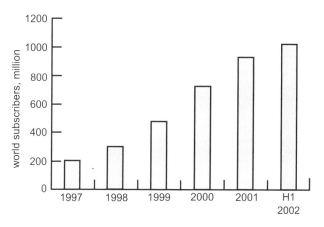

Figure 1.9 Number of world subscribers (millions) from 1997 to 2002

H1 = 1st half of 2002. Source: EMC database, **http://www.emc-database.com**

It must be noted that there is still significant growth potential in China with a population of over 1.2 billion (Yung-Stevens, 2002).

The industry is turning its eyes toward growing additional revenue streams from mobile data services, converting prepaid customers to postpaid accounts that traditionally have a higher ARPU and winning customers from a competitor's network to their own, known as churning. There is another feature of churning, which is referred to as handset churn, for example, where a customer typically ceases a contract after 12 months and signs a new 12-month contract and receives the latest handset at a subsidized

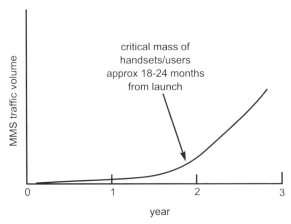

Figure 1.10   MMS growth pattern and device take-up

price highly drives the uptake of new handsets over an extended
period.

   The Asian market, in particular Korea and Japan, is the most
sophisticated in terms of handset availability and features. A
wide range of handsets have supported the Sha-Mail service from
J-Phone and more recently the i-mode service i-shot. From past
experience it has taken approximately two years for a market to
begin to mature, largely driven by the availability of devices that
support new services and the usability and usefulness of the service
(see Figure 1.10). This has been shown in the take-up of i-mode,
where from 1999 to 2001 subscribers migrated from voice-only
services to over 17 million subscribers using the i-mode data
services. Currently, in 2003 there are over 36 million i-mode
subscribers.

1.3.2 Convergence

The capabilities of mobile devices are increasing at a tremendous
rate; this is largely due to miniaturization of core component
technology and improvements in battery life and techniques to
conserve power. A typical specification for a Personal Digital
Assistant (PDA) in 2003 is 400 MHz with 64MB RAM and a built-
in Wireless LAN (Local Area Network) or GPRS (General Packet
Radio Service) connection; this is the sort of processor speed that
would have powered a desktop PC in 1997 and indeed for simple
word processing is still adequate. It is also interesting to compare

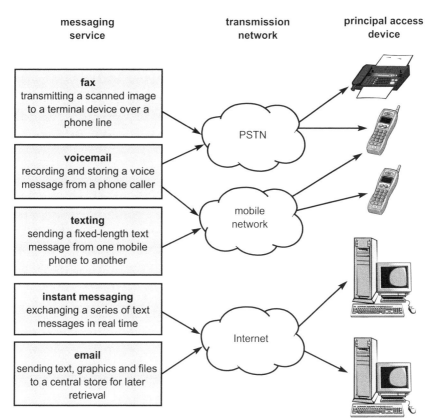

Figure 1.11   Basic messaging elements
PSTN = Public Switched Telephone Network

this with a Palm Pilot PDA from 1997 where the processor speed would have been 10 MHz and the memory footprint only 2 MB. This is leading to convergence, not just around fixed and mobile technology being replicated on a pocket device (although this is happening), but is more about convergence of service access: delivering the services used everyday, such as email, voicemail, fax and instant messages to any device, whether this is a fixed phone, desktop PC, PDA or other mobile device (Figure 1.11).

User interface challenges are very different with each access mechanism, but the driver is less about putting a technology in a pocket than delivering a service to customers while mobile. One technique to overcome the user interface challenge is the use of multimodal techniques with mobile devices; this is the use of voice to control services and input information into web forms.

While this is still an immature technology and the subject of research, there are a few venturing companies, exploring the commercial opportunities presented by the multimodal interface. Just a couple of the many products in this area can be obtained from Broca Networks **www.brocanetworks.com** and the Gabrielle System from BTexact **www.btexact.com**.

1.3.3   Toward third generation

There are significant evolutionary changes from circuit-switched networks that currently provide voice communication, the primary service while mobile. These changes are around the implementation of a packet-based network infrastructure to interconnect with the IP-based networks deployed in the fixed network. We will outline the migration from GSM data, through GPRS data to Wideband Code Division Multiple Access (WCDMA).

   The data rate for a GSM timeslot is 9.6kbps after error correction overheads, and this usually has a latency of 250 ms. It continues to be used for legacy handsets to access WAP-based applications or for specific applications that require a dedicated constant bit rate. The data rates for GPRS are quoted as up to 100kbps although in practice the range of available devices in 2003 support up to 30–40kbps. There are a number of additional factors that affect the data rate achieved with a GPRS connection:

- downlink/uplink ratio;

- number of users sharing the available channels;

- network dimensioning for GPRS data.

The downlink/uplink is often expressed as $3 + 1$ or $4 + 4$; this refers to the maximum data rate possible. Many mobile devices with limited battery life, such as phones, are of the type $3 + 1$, which indicates 30kbps down and 10kbps up. The reason for this is the power drain to transmit is much greater than when receiving, which is more passive. A few phones are capable of operating $3 + 2$ and $2 + 2$ to improve the upload data rate. The market for $4 + 4$ is a PC card format for use in laptops, where power consumption to transmit data is considered less of an issue, and in applications such as large email upload or video-conferencing. It must be noted that a maximum of five slots can be used in any configuration. This asymmetric data rate does not impact the transmission of picture messages as this is performed in the background, once the send button has been pressed.

The number of users sharing the available GPRS capacity and the network dimensioning for GPRS traffic have a significant effect on such services as video streaming, which requires a minimum bandwidth, normally of 20–30kbps for acceptable MPEG4-encoded video. The implications of video streaming on MMS is discussed further in Chapter 4. The effect of network dimensioning should not be underestimated as, currently, priority is given to voice traffic on available channels, usually with a minimum single channel allocated to GPRS data traffic. If all voice channels are in use, say at a football stadium, then available GPRS data may be limited to a single 10kbps channel which in turn may also be shared by other users.

A further factor that will affect multimodal service deployments in the future, as the transition is made to an all-IP network, is that of GPRS device classification:

Class A   This class includes devices that are capable of working with circuit-oriented and packet-oriented connections simultaneously. With these mobile telephones, users can begin transferring data during an active voice connection. Class A is at the high end of the GPRS mobile telephone spectrum and includes multifunctional devices or "smartphones."

Class B   This class includes mobile telephones that can support either a voice connection or a data connection. Trying to establish a simultaneous connection with both does not work. Switching between one service and the other happens automatically. The majority of GPRS mobile telephones fall within this category.

Class C   In this category, users must manually switch between GPRS data transfers and GSM voice connections. With type C devices, simultaneous service for both types of transmission is not possible. Devices in this class are for the most part oriented to PC-based applications or utilized for such applications as remote surveillance.

Various trials and commercial launches are under way for 3G networks around the world; these early launches are really a stepping
stone to test services deployed on early 3G and 2.5G (Second and a
Half Generation) networks that will enable a reality for the mass
market.

The improvements in frequency reuse and overall data capacity
provided by WCDMA will allow many more users to access data
services, such as video messaging within a particular cell. The theoretical 2 mbps data rate is unlikely to be offered as a commercial
service for a number of years as WCDMA is better suited for data
transmission between 64 kbps and 384 kbps. This will be adequate
for many users' needs, particularly as WCDMA supports dedicated
data channels to maintain quality of service.

Further proposals for using Session Initiation Protocol (SIP) on an
all-IP network are included in the 3GPP Release 5 standards; it
remains to be seen whether the business case for an all-IP
network is possible in the short term, but undoubtedly as a
vision a single IP network for voice, video and Quality of Service
(QoS)-enabled data services is a goal (Wisley et al., 2002).

Finally, the introduction of public Wireless LAN (WLAN), referred to as 802.11b, around the world is increasing in momentum;
a number of users can share the data rate available on a particular
access point. An access point normally has an omnidirectional
coverage of about 200 m; this can be extended with increased
power. A directional antenna can be used for point-to-point links
over several miles. The available data rate at the access point can
range from 2 mbps to 54 mbps depending on the coding scheme
and frequency in use. The number of users per access point is
theoretically unlimited; however, around 30 users who are not
employing intensive data transfer applications is recommended
and stacking access points can be used to alleviate dense traffic
demands within a specific area. Work is under way in 3GPP to
standardize the integration of WLAN and 3G networks, this will
begin to appear in Release 6 and proprietary implementations are
undergoing trials at present (2003).

1.3.4 Support technologies

As a standalone proposition MMS is unlikely to succeed as anything other than a replacement for SMS. The enhanced service

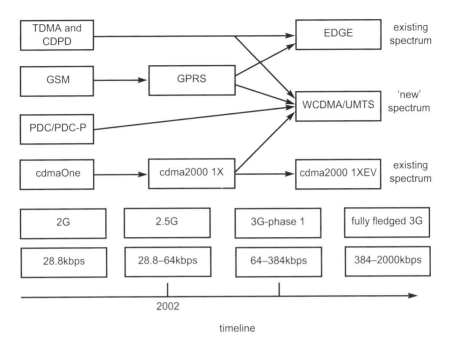

Figure 1.12 3G transition timeline
TDMA = Time Division Multiple Access; CDPD = Cellular Digital Packet Data; EDGE = Enhanced Data rate for GSM Extension; GSM = Global System for Mobiles; GPRS = General Packet Radio Service; WCDMA = Wideband Code Division Multiple Access; UMTS = Universal Mobile Telecommunication System; PDC-P = Packet Data Cellular-Protocol; 1XEV = a model number for Qualcomm cdma 2000 range

offering will be through the addition of cameras fitted to devices for picture messaging, animated messages from content providers and video messaging. Also, there are some key additional technologies that will be essential in supporting the successful adoption of MMS:

- location information;
- personalization;
- mobile commerce;
- unstructured supplementary services for data.

1.3.4.1 Location information

The 3GPP Release 5 architecture (TS 23.271) specifies the location client services components for retrieving location information for a mobile device in the network. Currently, the location information can be retrieved for a mobile terminal, assuming appropriate authentication, with an accuracy proportional to the radius of a particular mobile base station mast. Many masts are split into three sectors and may be 200 m in dense areas such as train stations, airports and up to 15 km in rural areas. The example below shows how an entry in the Redknee (**www.redknee.com**) implementation of a gateway mobile location centre may appear:

```
<property>
  <loginList>
    <login userid="test1" password="test1" />
    <login userid="test2" password="test2" />
  </loginList>

  <locInfoList>
    <locInfo msisdn="111111" locdata="83 11 22N|12 23
    4W|22" locage="3"retcode="0" />
    <locInfo msisdn="123456" locdata="84 25 5N|5 44
    7W|100" locage="2" retcode="0" />
  </locInfoList>

</property>
```

If an authenticated application were to query for the location information for Mobile Station ISDN (MS-ISDN) 111111 with a location age (locage) of no older than 5 minutes, then it would be returned the location data in latitude, longitude, radius and the age since the last update. Obviously, the demands of the application may not dictate any greater accuracy than is available. A typical scenario would be: if you are retrieving tourist information on a city or area, a map of the local points of interest could be delivered using MMS. Early examples are available today: for instance, J-Phone's J-Navi service. Analysys expects that 50% of all subscribers will use such services, with a global revenue of $18.5 billion by the end of 2006 (Wisley et al., 2002). Note that:

1. HSS includes both 2G-HLR and 3G-HLR functionality and LCS is included in the overall network architecture in TS 23.002 (Figure 1.13);

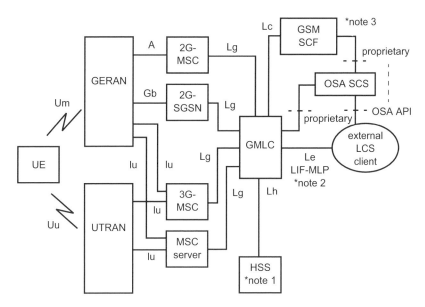

Figure 1.13   3GPP Rel 5 LCS architecture

3GPP = Third Generation Partnership Project; UE = User Equipment; GERAN = GSM/EDGE Radio Access Network; GSM = Global System for Mobile; EDGE = Enhanced Data rate for GSM Extension; UTRAN = UMTS Terrestrial Radio Access Network; UMTS = Universal Mobile Telecommunications System; 2G-MSC = Second Generation-Mobile Switching Centre; SGSN = Serving GPRS Support Node; GPRS = General Packet Radio Service; GMLC = Gateway Mobile Location Centre; HSS = Home Subscriber Server; SCF = Service Compatibility Features; LIF-MLP = Location Interoperability Forum-Mobile Location Protocol; OSA SCS = Open Services Architecture Service Capability Server; LCS = Location Client Services; API = Application Programming Interface

2. Location Interoperability Forum (LIF)–MLP may be used on the Le interface;

3. in one option the LCS client may get location information directly from GMLC, which may contain Open Services Architecture (OSA) mobility Service Capability Servers (SCSs) with support for the OSA user location interfaces (see 3GPP, 2002c, f).

1.3.4.2   Personalization

Personalization is formed from the concept of a user profile and has received considerable attention as a means of attracting and locking users into particular services (Ralph and Searby, 2003). Mobile telephones are very personal devices. They can hold an individual's personal contact information, they fit in the pocket and they are taken almost everywhere. The frequency with which people customize the appearance and the sound of the mobile provide further evidence that they are treated as a personal fashion statement. Functionally, as well, they provide a communications medium for contacting a person rather than a location (a desk or a house) or an organization.

Personalization applications can currently be grouped into four areas. This provides a means of identifying the value proposition and makes it easier to separate out some of the technology market sectors:

- targeted marketing/advertising;

- customer relationships;

- service integration;

- knowledge management.

1.3.4.2.1 Targeted marketing/advertising

The area of targeted marketing/advertising is centred on generating increased revenue mainly in the consumer market. It is widely recognized that the Internet has shifted the power base toward the consumer. The consumer can access many suppliers to get the best deal and, with such a wide choice available to potential customers, it is now becoming necessary to do more to retain their attention. That is not to say that personalization techniques could restrict choice (although this is possible if the only index and search capabilities accessible to a user were under the control of individual service providers – as might occur on a digital TV network), but they could allow the shopping experience to be less frustrating and to tempt customers with goods that are most likely to interest them.

Personalization has the capability to create the impression of delivering a niche product to each user.

1.3.4.2.2 Customer relationships

The term "Customer Relationship Management" (CRM) has often been used in the context of personalization. It represents more than just a technology approach: it is a change of business philosophy. Rather than building an organizational structure and business processes around individual product or service groupings, a customer-centric approach is taken. This is intended to provide improve-

ments in customer experience so that he or she is always recognized
as the same individual regardless of the channel used to contact the
organization. In addition, system integration is applied to ensure
that all communications channels and operational systems link to a
common user account profile. Consequently, as well as driving
toward greater customer loyalty, the provision of multiple communications channels, including online access, can drive significant
cost reductions in the provision of customer support. For cost reductions to be really effective the quality of the user experience
through automated systems needs to be excellent; the ability to
interface through a range of devices and in multiple languages
needs to be considered as well.

1.3.4.2.3   Service integration

A symptom of the drive to launch new products and services
quickly, combined with organizational structures centred around
products and services, has been the implementation of systems in
isolation of the existing infrastructure. These often have their own
customer database and single-access channel.

Situations can arise where a call centre Interactive Voice Response (IVR) system may have no integration with an online
knowledge base despite supplying the same content from the
same organization. Overcoming this kind of legacy problem can
prove extremely costly, as it is likely to lead to large system integration activities or replacement with unified service platforms.
The use of personalization technology can offer a lighter weight
solution in many cases.

1.3.4.2.4   Knowledge management

Trends in the business environment are placing enormous pressure
on users to manage an ever-increasing workload. Business users'
demands to access information while away from the office, and to
process it, will drive the emergence of services to filter and target
information to match users' interests and the particular task in
hand. Some of the key features in this area include:

- responding to user context;

- matching information to interests and preferences;

- sharing information with other appropriate users in a
  community.

User context can include the geographic location, the type of device and access mechanism in use and users' current role (e.g., at work, at home). Matching to both interests and context provides clear benefits when searching for information. Indeed, the interests and context can be very dynamic and may need to be adapted many times during a typical day as users switch between projects and activities. Community tools are becoming increasingly important to allow people to collaborate more efficiently – especially when their working environment is spread over multiple locations. This is not just relevant in a business context, though, as there are opportunities for consumer services offering collaboration online, and many claim to be generating significant revenues from their membership.

1.3.4.3 Mobile commerce

Mobile commerce is placing greater importance on the mobile phone. As a device enabling secure transactions such as banking, payment and ticketing, the mobile phone is becoming a Personal Trusted Device (PTD). New personal features in mobile phones will make PTDs the ultimate digital wallets, which will be used for a wide range of mobile transactions either remotely, over the digital mobile network, or locally at a point of sale. Mobile commerce has started with remote transactions in the mobile Internet. Soon, concepts like smart covers (where a smartcard chip is embedded into a replacement phone casing) will facilitate the development of local transactions in a fast, easy-to-use and secure way. This opens up huge opportunities for digital wallets.

In addition to buying physical goods, a considerable part of mobile commerce consists of the purchase of different types of digital content delivered using MMS that in most cases ends up being utilized in the mobile phone. Consumers want to personalize their mobile phones with ringing tones, graphics and picture messages from content providers. Games, downloadable phone applications as well as music and video feeds will soon follow. Many of these purchasing scenarios are enabled by sending an SMS request to a shortcode that charges the phone bill with an amount up to £5. Dialling a premium rate phone number, where the call is answered by an automated service, is also used success-fully for charging for content. Vodafone offer the M-PAY service to allow users to select content to be sent to a mobile device, which is charged directly to the phone bill.

Key drivers for mobile commerce service adoption will be ease of use and security. Too complex and time-consuming applications and services create barriers that may discourage consumers from going "mobile". The challenge is to implement the secure payment scheme so that it remains convenient and simple to use.

1.3.4.4   Unstructured Supplementary Service Data

Lastly, Unstructured Supplementary Services Data (USSD) is an asset in the GSM networks today, but one that is not very often exploited! In some networks USSD is used for loading of prepaid accounts, etc.; but it can be used for a lot more. USSD is a session-oriented protocol suitable for interactive, menu-driven sessions, where a subscriber requests text/data from the system. Through the USSD gateway requests/commands can be sent, under full operator control, to numerous sources of information, operator internal systems or to the Internet. Mobile terminals of phase 1 and 2 are ready for USSD. System-originated, or pushed, USSD messages require phase 2 terminals.

A USSD message can be a maximum of 182 characters long (i.e., comparable with SMS). Delivery of the first USSD request and response is comparable in response time with an SMS. However, the following USSD messages in a dialogue are delivered significantly quicker than SMS. This is achieved because, when a session is initiated, the radio connection stays open until either the terminal or the system releases it. A USSD session typically starts with the subscriber sending a USSD message similar to *100# SEND. The message is sent through the GSM network to the USSD gateway (Figure 1.14), which interprets the code to a specific request for information from a defined source. The information source can be a hyperlink to an Internet site, or information stored locally in the service provider's domain. After retrieval, the information is converted and sent from the USSD gateway into the network and to the mobile terminal. Other services in operation based on USSD are:

- prepaid balance check;

- voucher refill;

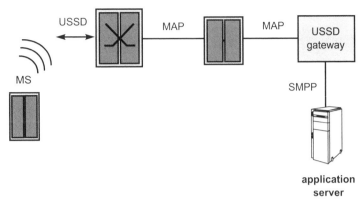

Figure 1.14 USSD architecture

MS = Mobile Station; USSD = Unstructured Supplementary Services Data; MSC = Mobile Switching
Centre; MAP = Mobile Application Part; SMPP = Short Message Peer to Peer

- activate WAP service;
- activate MMS service;
- configure MMS settings OTA;
- activate and manage family and friends;
- vending machine ticketing;
- pay for tickets in public transport;
- gateway to WWW information.

1.4 THE CHALLENGES OF MULTIMEDIA MESSAGING

The introduction of the MMS is a significant milestone in the evolution to 3G services. Following the launch of MMS will be the introduction of unique services, usage scenarios, business models and regulatory challenges. In this section we summarize some of the key factors to impact the success of MMSs.

1.4.1 Usability

The importance of customer experience in using a service should not be underestimated. The poor usability of WAP services generated a lack of interest in the services offered, especially when coupled with a charging mechanism based on connection time.

The device format for many MMS users will not change significantly from that used for early WAP services: there may be the addition of a small joystick for menu navigation and slightly larger colour screens with better resolution. Many devices include MMS as an additional messaging feature alongside SMS or email. The danger is that users may consider MMS as a simple replacement to SMS, offering little more than is possible with email.

Usability extends through the user interface and signifies how users, particular those who just want to use a service, can be guided through a service such as MMS without the need to understand the technology beneath. A customized user interface is at the heart of the Vodafone Live! offering. This user interface is delivered across a range of devices, and offers a consistent menu system. One such technology that supports and enables this level of customization is the Trigenix user interface development environment from 3Glab (Davies, 2002). Trigenix is made up of three components centred around the use, design and delivery of graphics, sound and animation packages, including OTA deployment. A compact application resides on the handset and enables interaction between the handset's software components (phone, messaging, contacts, calendar, etc.) and the user interface. Operators can use this to drive revenues by branding the user interface, dynamically promoting services on the screen and providing seamless connection, making it easier for users to buy services.

As mentioned earlier, another solution that does not require the devices to be made physically larger is to add voice control and audio feedback to the traditional mobile phone interface to create a *multimodal* interface. Using speech recognition technology, navigation through an application becomes easy with voice commands, while data entry is made considerably simpler. Audio feedback can be used to supplement what can be shown on the limited screen space. Speech recognition, however, is computationally expensive, and the more complex recognition tasks are beyond the capabilities of current MMS devices. Distributed processing solutions are therefore required to deliver the full range of multimodal-enabled applications, with those speech recognition tasks that are demanding carried out on network-based servers.

We will see as we move further through the chapters how MMS can be used for video or animation and integrated into premium content services. Its ease of use and seamless access will impact its uptake and profitability.

1.4.2 New usage scenarios

MM and improved device capability will offer a number of services and applications that would not be possible from a desktop PC: apart from simple, novel ways of keeping in touch by sending picture messages, receiving a reminder for a cinema ticket you have purchased, or a notification that a friend is nearby, you could look up a map for the nearest pizza restaurant or buy a coke by scanning a voucher from your mobile phone that was sent using MMS.

Arguably, it is the convenience and immediacy that will drive consumers to access services based around MMS. Identifying these time-critical tasks and developing them into compelling services with profitable business models will unlock its potential. Emerging usage scenarios are a key theme of this book (Figure 1.15). Chapter 7 includes a brief discussion of where MMS may be heading, based on insight from ongoing development projects.

The Work Foundation in the UK is an example of a project with a remit to assess the impact of technology in the i-society. The social context in which mobiles are used will have to be taken into

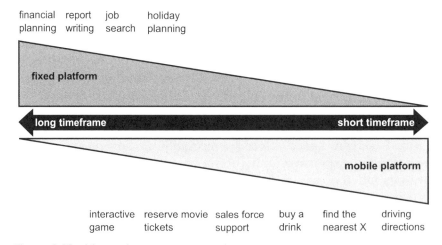

Figure 1.15 Messaging usage scenarios

account if future services are to succeed. This is the conclusion of a report the Work Foundation's i-society project, which looks at how technologies such as broadband and mobile usage impact people's everyday lives. The authors of the report spent time with four families to find out how people interact with their mobiles. The results make sobering reading for phone companies on the verge of launching expensive, data-rich mobile services (**www.theworkfoundation.com/research/isociety/MobileUK_ main.jsp**).

1.4.3 New business models

There are many players moving into the mobile messaging space, but the mobile network operators dominate due to the ownership of customer billing information (see Figure 1.16). Handset manufacturers, particularly Nokia (with a 39% market share in 2002), have strong brand recognition with the consumer. Both these factors cannot and do not ignore the fact that content is the important element of successful messaging services. To help encourage market acceptance for more sophisticated pricing, Radiolinja is considering separating voice billing from entertainment billing by evolving pricing for mobile media, coupled with distinct billing and enhanced services, in preparation of the Finnish market for 3G. This approach has been adopted by Hutchinson 3G (UK) (**www.three.co.uk**, and the use of price bands for specific data services or content downloads is attributed against the entertainment value bundle that a user has subcribed to.

Vodafone pays up to 60% of premium revenues from Vodafone Live! services to content providers. Nokia is constantly promoting service development with the Nokia Forum community (**www. nokiaforum.com**), as indeed the UK operator mmO$_2$ does with an Application Development Forum (**www.sourceo2.com**).

MMS uses proven billing models within mobile networks, such as revenue sharing, subscription and sponsorship models. However, the use of "shortcodes" for specific tariffs associated with charging mechanisms for "premium content" delivered by SMS is to be replaced, by rating and mediation of content services. This will allow specific types of content to be billed, without the need to remember a shortcode. It also requires support for "advice of charge" mechanisms, as discussed in Chapter 5. Supporting and

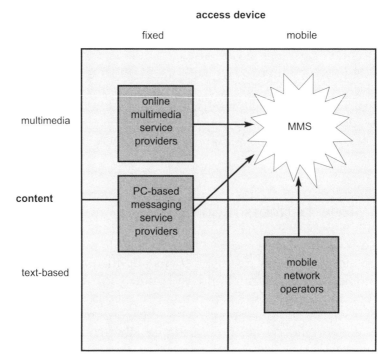

Figure 1.16 MMS collision

encouraging the user to access services with transparent pricing models will be vital; the alternative is that it alienates the user.

The association of shortcodes to fixed pricing tarriffs has made SMS pricing complex to implement for service providers, as they need to differentiate between services by either using different shortcodes or asking the user to enter keywords in the message to ensure it is processed correctly. The situation is not likely to improve as users start to send and retrieve protected as well as unprotected pieces of multimedia content across mobile networks. Operators face a huge challenge to get end-user acceptance of their mobile data pricing. The launch of MMS services will in many respects bring even further complexity to the already difficult pricing structures.

It is important not to underestimate the fact that the introduction of these business models needs to meet consumer expectations of value for money. There remain major obstacles that can prevent MMS from being a commercial success. As mentioned earlier (p. 10), with voice revenue per user declining for mobile network operators, they hope to increase revenues on other services; in the UK many mobile operators are targeting increasing data revenue from 10% to 25% by 2004–2005. MMS is clearly an important part of this strategy and must be managed so as not to simply cannabalize existing SMS revenue. We will discuss this specific threat further (Northstream, 2002a). The roaming nature of the GSM world has led the wireless industry to identify at an early stage the need for a common approach regarding MMS pricing. To help the situation the GSM Association's Billing, Accounting and Roaming Group (GSMA-BARG) has issued detailed recommendations for how MMS content could, and preferably should, be priced. If common principles are accepted internationally it will form a good basis for quick and easy set-up of crucial roaming agreements. The proposed principles of the GSMA are:

- sender pays for MMS (calling party Pays—recipient does not pay);

- integrated event charge per message (i.e., the price of MMS includes the content and the associated bearer);

- level of event charge, based on volume bands.

Telenor, the first company to launch MMS services commercially (in March 2002) charged its end-users €1.30 per sent message (at the time of writing), irrespective of size. This is not in line with GSMA recommendations, which prescribe a price per message, based on which predefined size category the message falls into. To be able to follow the pricing principles suggested by GSMA, with a predefined fixed price for messages of a certain size, there must be billing network elements in place to process the Charging Data Records (CDRs) of MMS data from other types of down-loaded data. There are several ways to implement the required functionality into networks, but progress is slowly beginning to be made in communicating the mobile operator's price structure to the end-user. GSMA has identified the issue in its recommendations and has proposed a model for a mechanism in terminals that would indicate the price category each message belongs to (known as "advice of charge").

   It should be noted that every MM sent requires an SMS notification (using WAP-PUSH) to enable delivery. It follows that if SMS traffic is migrated to MMS then 50% of SMS traffic will be maintained in a mobile network. It is with this in mind that mobile operators need to ensure high-capacity SMS network elements continue to operate, and, ultimately, while the operator cost of an SMS is 1p or less (due to such high volumes), this must still be recovered within the MMS charges.

Vodafone UK has released a tariff (early 2003) of 15p for a 1.5KB text-only MMS. This will aim to ensure a smooth migration from the text-only SMS services to MMS. However, most users regard both as simply messaging services and argue they should not be penalized by a high charge for what is the same functionality. If operators prove unable to provide coherent and transparent pricing schemes for their messaging services, they expose themselves to the risk of revenue cannibalization from upcoming messaging technologies with replacement potential, such as Post Office Protocol version 3 (POP3) email or Java messaging applications.

The packet-based nature of GPRS makes it suitable for IM services. IM has proven to be a killer app for the fixed Internet and fills a similar basic need as SMS, sending simple and fairly unobtrusive text messages between people. According to Ferris Research it is projected that the IM market will grow from 10 million to 182 million users by 2007 (Ferris Research, 2003). At present, no mobile IM client has been able to match ease of use presented by a normal mobile phone's SMS client, but this could change quickly. Java or C (Symbian) could be used to develop an IM client. Already, third-party applications can be downloaded and run on the mobile terminal. The Microsoft SmartPhone OS also includes a mobile version of the MSN messenger IM client.

Phones with this capability are becoming available and any IM provider could get client software installed on each and every phone. Suddenly, IM could be a viable alternative to SMS, if the price vs. ease-of-use equation is unbalanced. There is an obvious potential threat to operator revenues from such a development. If a text message (transmitted over IP using GPRS) contains 65 characters on average and if half of the transferred and billed data is protocol overhead, a message would cost £0.002 (using 1KB = £0.02).

The discussion is continued in Chapter 2 where specific business models are considered.

1.4.4 The interoperability challenge

Due to the evolutionary and revolutionary nature of MMS it was inevitable that the initial services have suffered interoperability issues. The complexity of the value chain and consumer demand for services across countries and networks at affordable prices require network interoperability to achieve a sustainable business model. While MMS is built on technical specifications from 3GPP it has a number of areas that are either undefined or implementation-dependent. These range from billing and content provision to device presentation. This lack of clarification leads vendors to implementation decisions that may introduce incompatibilities and interoperability issues.

The Open Mobile Alliance (OMA) has worked with several vendors to produce the MMS conformance document, which adds further clarification to details of the specification. While MMS offers new opportunities, there are still some interoperability obstacles and barriers to be overcome:

- *Device compatibility* – many services require you to specify your phone before downloading MMS content as it is important to check compatibility with the device. Some content types are incompatible or unsupported on different devices.

- *Device configuration* – OTA programming is offered by mobile operators to enable users to automatically configure their mobile phones, GPRS and MMS settings by receiving an SMS. But once again there are vendor-dependent elements to OTA programming, which means that not all phones can be configured in this way. Website help pages providing configuration information and help are essential to avoid call centres being flooded.

- *Network interconnect* – the MM4 interface supports the interoperability of interoperator exchange of MMs. The commercial and technical obstacles are discussed in Chapter 5.

Throughout this book, we will discuss a variety of problems and make efforts to address them.

1.4.5 Security and privacy

Security and privacy are important areas where multimedia messaging will break new barriers. The question of security will not be
addressed in terms of encryption of the air interface but in the
digital rights management and security of content, once delivered
to the device. This is interwoven with the security issues surrounding billing and managing premium content downloads. However,
one of the most challenging aspects of multimedia messaging is
how to find solutions and services that reconcile the users'
demand for highly personalized services with their desire for
privacy. Organizations such as the World Wide Web Consortium
(W3C) and the Mobile Marketing Association (MMA) are attempting to define standards that will help preserve the privacy of mobile
users and go some way to addressing the issue of spamming.

2

The Multimedia Messaging Value Chain

2.1 INTRODUCTION

Before we delve into the technologies, applications and services of multimedia messaging, it helps to have an overall view of the value chain. As we saw in Chapter 1, multimedia messaging needs a variety of business partnerships often involving a number of organizations: from content providers, mobile network operators to infrastructure providers. The objective of this chapter is to introduce you to some of the main categories of players involved in the creation and delivery of Multimedia Messaging Services (MMSs), to help you understand the role they play and the context within which they operate. This chapter also provides a discussion of the primary business models and concludes with an outline of digital rights management.

The delivery of MMSs involves a number of players. They generally fall into one or more of the following categories (see Figure 2.5 on p. 47):

- infrastructure equipment vendors;

- content providers, including application/service developers;

- mobile device manufacturers;

- mobile network operators.

Multimedia Messaging Service   Daniel Ralph and Paul Graham
© 2004 John Wiley & Sons, Ltd   ISBN: 0-470-86116-9

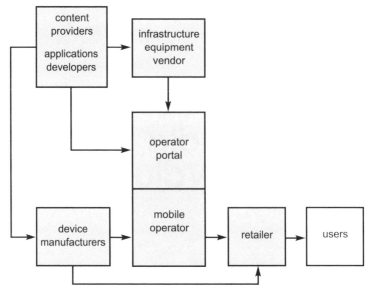

Figure 2.1 MMS value chain

2.2 INFRASTRUCTURE EQUIPMENT VENDORS

The traditional mobile network infrastructure vendors such as Nokia and Ericsson have secured a number of the Multimedia Messaging Services Centre (MMSC) contracts awarded by mobile operators around the world. While these are considered dominant players in the circuit-switched domain, they are competing and forming partnerships with new players who approached mobile telecoms from the Internet domain with the deployment of Internet Protocol (IP)-based email messaging services. The following two categories broadly represent these:

1. *MMS centres* – network vendors and systems integrators are marketing MMSCs, the basic network element needed to provide an MMS service. Ericsson, Nokia, LogicaCMG and Comverse are among the leading MMSC vendors.

2. *Messaging platforms and software* – the vendors of messaging platforms and software are entering the MMS market at a rapidly increasing rate. Leading messaging vendors with MMS products include Materna, Openwave, Tecnomen, CriticalPath and Tornado.

Other vendors are providing specialized network elements for MMS solutions, including billing systems from companies such as Argent, VoluBill, Xacct, InTec, NEC and SchlumbergerSema and imaging servers from companies such as LightSurf and ArcSoft. As an example of one of the infrastructure providers' products, ConVisual supply a multimedia messaging broker, which is a platform for the creation, personalization, conversion and distribution of multimedia messages (MMs), supporting applications such as "visual info services", "mobile advertisements" and "picture postcards".

As major stakeholders, infrastructure vendors are also playing a critical role in key standardization and interoperability initiatives, including:

- The Third Generation Partnership Projects (3GPP and 3GPP2), which aim at developing global specifications for Third Generation (3G) mobile systems based on evolutions of Global System for Mobile communications (GSM) and Code Division Multiple Access (CDMA) technologies, respectively.

- The Open Mobile Alliance (OMA), formerly known as the Wireless Application Protocol (WAP) forum, which works on the development of mobile internet standards. This includes the Location Interoperability Forum (LIF) and Wireless Village fora.

- The World Wide Web Consortium (W3C) manages the development of the specification for the Synchronized Multimedia Integration Language.

- The Internet Engineering Task Force (IETF) manages the development of specifications for Simple Mail Transfer Protocol (SMTP) and Multipurpose Internet Email Extension (MIME), among others, using a process and naming convention called Request For Comments (RFC).

2.3 CONTENT PROVIDERS

Above all, multimedia messaging is about content and giving users access to additional services. Content can range from news updates, directory services, directions, shopping and ticketing services, entertainment services, financial services and so on.

Companies such as FunMail, ExtraTainment and Newbay are developing innovative MMS service applications (e.g., Customer Relationship Management [CRM] follow-up by MMS to say "thank you", the cartoon strip "Recycle Dog" and personal publishing).

An MMS application development company, Iomo, has launched Foto FunPack, which allows MMS content to be manipulated with the addition of "joke" backgrounds or be used as the picture for a "sliding tile" puzzle game. Nokia lists a number of different third-party solution providers for a range of market sectors, from corporate, consumer and operator to government (**http://www.nokia. com/nokia/0,5184,1849,00.html**). Major television content providers are also showing interest in MMS opportunities: perhaps the most notable example to date is the announcement by Nokia in the first quarter of 2002 of MMS partnerships with the BBC, Euro-sport, Lycos and MTV.

In the promotional white papers and video clips heralding MMS, vendors such as LogicaCMG and Ericsson present a world in which people can spontaneously take snapshots with the cameras built into their MMS phones and immediately send them to friends, family and colleagues. This self-created multimedia content will, in time, be a large percentage of MMS traffic. In order to encourage growth in the MMS market, it will be necessary to provide access to a large, varied and attractive range of ready-made content that users can add to their messages. Even with camera-equipped handsets becoming more widespread, an excellent range of ready-made content will still be an important means for MMS service providers to differentiate themselves from competitors (Strategy Analytics, 2002).

Users who add their own pictures to their messages will likely want cartoons, greetings and other types of ready-made content. This is based on the tremendous growth in the Japanese market for these content types, but it should be noted that while this technology may break new ground it is unlikely to replace the traditional greeting card.

SonyEricsson have launched in 2003 a content album website called "My MMS" and a selection of downloadable MMS content in an "MMS gallery". This will help the industry by seeding the infrastructure required to allow content to be stored and retrieved after its creation or initial download. This is a significant difference to Short Message Services (SMSs), where only a ringtone or logo was downloaded to the phone and stored. Now potentially many different and wider ranges of content can be downloaded and stored for retrieval later (**http://www.sonyericsson.com/fun/**).

Many services will not have requirements for content storage or be suitable to the transient and timely delivery of content, such as a traffic map or an advert for a same-day offer. So, for Person-to-Person (P2P) MMS services, ready-made content will be a very important element. For application services based on MMS (e.g., Machine-to-Person [M2P] services), ready-made content will be an even more important element. A service that sends a message to your phone when your football team scores, for instance, would be much more exciting if it also sent you a videoclip of the goal being scored. MMS gives scope for a wider and more diverse range of applications than those that are possible with SMS.

2.4 MOBILE DEVICE MANUFACTURERS

Device manufacturers are an important part of the multimedia messaging value chain. They are the suppliers of the operating systems, microbrowsers, MMS clients, video players, email systems and other middleware technologies.

There is an ever-wider range of devices on the market, whose expansion to include new, innovative, device formats and licensing of core operating systems has enabled new players to enter the market, leaving aside the design decisions made in determining such new devices as General Packet Radio Service (GPRS) or Bluetooth capability, Wireless Local Area Network (WLAN) support and processor speed.

The device format is being adapted to the application. For instance, Nokia has released a device for Gaming, with a keypad either side of a central screen, and prototype wristwatch Personal Digital Assistant (PDA) devices are commonplace. A device called Cybiko (Figure 2.2) has been on the market since 1999. It does not require a subscription and cannot transmit voice, but can transmit and receive proprietary-formatted text messages in a 100-m radius of other Cybiko users. It does not use GSM or CDMA-related standards, and so cannot communicate directly with these handsets. All this highlights the different ways multimedia messages can be delivered to a range of handsets. However, the requirements for

Figure 2.2 Cybiko
Source: Cybiko website

each in terms of capability and presentation vary greatly and cause interoperability issues.

Handset manufacturers are driving the integration of new software products onto next generation devices, which inevitably require interoperability testing to ensure standards conformance. Each vendor is looking for commercial advantage through value differentiation. This tends to be exploited through intellectual property inherent in the software. An example may be a video player that starts with a lower frame rate while buffering a media stream, or the usability testing that has accompanied the release of an operating system (OS). In the operating system arena, the primary players are:

- EPOC (a small-device OS), developed by the Symbian Consortium, including vendors such as Nokia, Psion, Ericsson, Motorola, Matshustia and Samsung;

- Windows SmartPhone OS, specifically developed by Microsoft for mobile phones;

- PalmOS, which currently has around 50% of market share, having lost ground to Microsoft PocketPC 2002 edition OS for PDA.

Of particular note, the latest release – the Symbian v7 (Figure 2.3) platform – enables support for the JavaPhone API (Application Programming Interface) (address book, calendar, user profile),

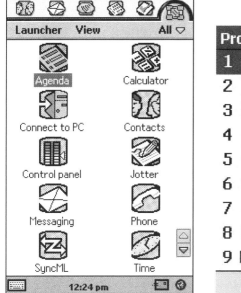

Figure 2.3   Symbian OS v7.0 and Microsoft SmartPhone
Source: Symbian website and Microsoft website, **www.microsoft.com**

IPv6, Hyper Text Transfer Protocol (HTTP) stack and web browser, SyncML and USB.

The market for video players will likely be more limited given the dominance (and in Packetvideo's case part ownership) of the mobile handset manufacturers:

- Packetvideo, the chosen Motion Picture Experts Group layer 4 (MPEG4) client for SonyEricsson mobile devices;

- Realnetworks, Nokia client for MPEG4 video streaming;

- Microsoft video player is to be used on Motorola handsets.

In Chapter 4 we will examine in more detail the range of terminals available and explore solutions for device interoperability.

2.5   MOBILE NETWORK OPERATORS

With falling profit margins in the mature mobile voice markets of Asia, Europe and North America, mobile network operators are

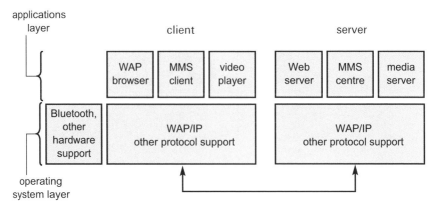

Figure 2.4 Block diagram outlining relationships between OS and applications

WAP = Wireless Application Protocol; MMS = Multimedia Messaging Service; IP = Internet Protocol

under increasing pressure to turn to mobile data services for additional sources of revenue. In the process, they are transforming from traditional voice carriers to multiservice providers that encompass wider segments of the value chain. The major components of the value chain in relation to an operator are:

• mobile messaging provider;

• mobile location wholesale;

• mobile portal;

• mobile billing provider.

Mobile operators in Japan and South Korea have launched proprietary MMSs since 2000, while most Western European operators launched MMS-based services in 2002 (Yung-Stevens, 2002) (see Figure 2.6). Operators in South Korea and Japan are planning to migrate their existing, proprietary photo and video messaging services to the MMS standard. Operators elsewhere in Asia are also very interested in MMS: SingTel Mobile launched services in the second half of 2002 and Singapore's other operator, M1, launched in August 2002. In South Africa the operator MTN launched MMS in 2002. In North America, AT&T Wireless is adding MMS services such as dating, greeting cards, weather, stock alerts and news during 2003. At the start of 2003 over 100 mobile operators had

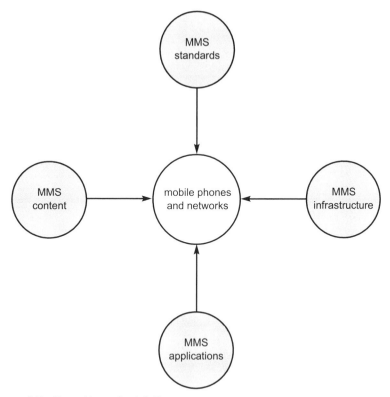

Figure 2.5 The drivers for MMSs

launched MMS and several third-party MMSC platforms were in operation (**www.mobile-metrix.com/press260203.html**).

Mobile operators are in a commanding position to provide users with a simplified, common, user interface, navigating services via simple menus and leveraging multiple sales channels including MMS, SMS, WAP and Java application distribution. Such mobile portals as Vodafone Live! are the starting point for offering their own and third-party MMS content and services, and as the market matures opportunities for service providers may be through Mobile Virtual Network Operator (MVNO) agreements. Here are a few simple guidelines to support the deployment of MMS:

- New content should be introduced (and advertised) regularly, and existing content retired or rested. This is necessary in order to avoid the same old pictures appearing time and time again in users' messages.

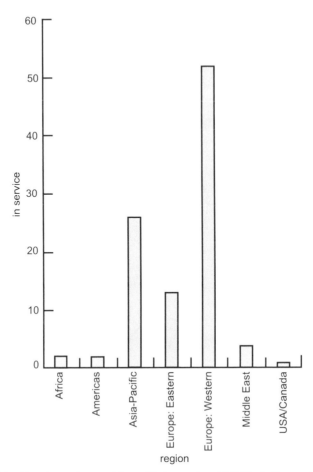

Figure 2.6 Number of countries with launched MMS platforms (2002)
Source: Nokia website

- The range should include both original and branded content – each has a different kind of appeal.

- The range should include more generally expressive content ("Hello", "Wish you were here" and so on) than occasion-specific content ("Happy birthday" or "Well done"). Operators must encourage everyday MMS usage, to avoid seeing it become something that is only used on special occasions.

- Concentrate on a core set of services supported by MMS, such as tourist maps, cartoons, news, traffic and sport information.

Interoperability between operators and message exchange between both operators and service providers will be vital in order to replicate the success of SMS with MMS. Operators have begun to negotiate national interconnection agreements, and many have successfully negotiated interconnect agreements around the world: Singapore operators signed the world's first interconnect agreement in November 2002, O_2 and Vodafone (UK) signed a deal in March 2003 and TeliaSonera announced interconnect agreements with operators in April 2003. We will discuss the technical billing arrangements of MM4 further in Chapter 5.

MMs can vary greatly in size and content (up to 300KB). To the mobile network, all SMS messages are identical: they all contain text only and are the same size (maximum 160 bytes). This helps to make traffic management (of the SS7 signalling links and network elements), interconnect agreements between operators and end-user pricing simpler for SMS. Successful operator interconnect will be one of the most important preconditions for high rates of MMS growth. The interconnection of operators' SMS services is widely recognized as being the major usage driver for the service (Oftel, 2002). For example:

- In the UK, the number of SMS messages for all mobile operators reached 599 million in the third quarter of 1999, before interconnect agreements had been established; since then, by the first quarter of 2002, this had ballooned to 4,136 million (Oftel, 2002).

- Wind, the Italian operator, registered a total of 50 million SMS messages on Christmas Eve, 2002. The number of MMS messages sent by Wind's customers on New Year's Eve of the same year reached 500,000 (**www.mobilecommerceworld.com**).

Interoperability will also be required at a service and device level to ensure that as far as possible users can receive content in the same or if necessary an adapted format from what was sent. We will discuss this further in Chapter 4.

One threat to operator dominance may be from such service providers as Zidango (**www.zidango.com**). The MMS market will be more open than operators are currently used to competing in. This is because MMS can interwork with PC-based email and instant messaging, possibly further constricting the business model for MMS, which will rely on high volumes to achieve low-priced

	2002	2003	2004	2005	2006	2007
SMS	231,517	484,298	775,260	966,151	984,834	886,877
Messaging (P2P)	205,494	425,581	668,214	798,348	792,114	687,423
Content deliver (M2P)	25,651	56,352	102,747	161,891	184,715	190,452
Requests and notifications (P2M)	372	2,365	4,299	5,912	8,005	9,002
EMS	2,451	18,779	66,237	122,957	169,458	170,996
Messaging (P2P)	2,122	16,636	58,776	105,848	142,367	139,410
Content delivery (M2P)	329	2,143	7,461	17,109	27,091	31,586
MMS	8,637	27,187	70,833	182,811	354,900	581,190
Messaging (P2P)	8,009	21,713	49,768	123,251	236,752	397,770
Content delivery (M2P)	626	5,238	20,133	56,455	111,884	173,054
Requests and notifications (P2M)	2	236	932	3,105	6,264	10,366

Figure 2.7 MMS traffic forecasts (millions of messages worldwide)
SMS = Short Message Service; P2P = Person to Person; M2P = Machine to Person; P2M = Person to Machine; EMS = Enhanced Message Service. Source: Delaney et al., 2002

services. As mobile messaging evolves to encompass non-text content, third-party content providers and the services built around this content will become increasingly important players in the service value chain (Delaney et al., 2002).

With mobile voice services, the user's perception of "the service" has until recently been centred on pricing, the (small) size of the phone and the network coverage to support it – all within the domain of the mobile operator. With mobile information, entertainment and commerce services, this perception will often be dominated by the content and applications, as shown in Figure 2.8, while the phone and the network are simply regarded as the means of access.

The move in importance toward the role of the service provider has the potential to erode the customer relationship with the mobile operator. Not withstanding, the operators monthly bill will be a primary means of contact with its customers, and this will be closely guarded by the operator. Some subscribers may use sports websites or familiar portals such as the Micro Soft Network (MSN) to access their mobile messaging services, especially if this is an extension of an existing mail service they currently use.

Service providers must enter the market with seamless, convenient services that fit with the way users currently interact with them. They must not underestimate the strength of the relationship between an operator and its customer, and the control this offers over the way they access services and the billing mechanisms

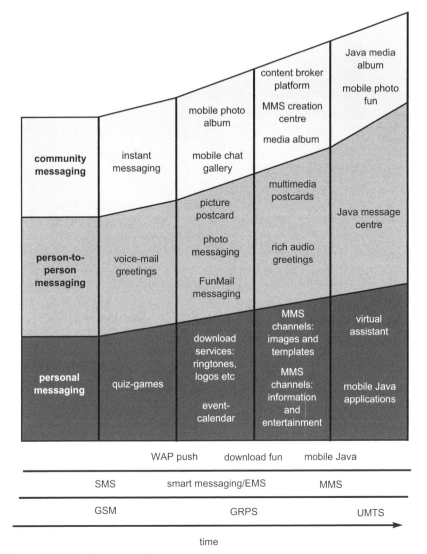

Figure 2.8   Services roadmap

MMS = Multimedia Messaging Service; WAP = Wireless Application Protocol; SMS = Short Message Service; EMS = Enhanced Message Service; GSM = Global System for Mobiles; GRPS = General Packet Radio Service; UMTS = Universal Mobile Telecommunications System

that can be employed, such as "free" trial periods of WAP browsing and "free" MMS for a month. Primary business models include revenue share and various forms of sponsorship, such as advertising, anchor tenancy, referral fees and commissions. Subscription-based business models are valid, but considered constraining by many users because much of the MMS traffic generated will be billed on a per-message basis. Digital Rights Management (DRM) will fundamentally support the validity of these business models.

The implementation of billing models and architectures will be discussed in Chapter 5. This section discusses the following in relation to the content provider:

- revenue share business models;
- sponsorship business models;
- subscription business models;
- DRM.

2.5.1   Revenue share business models

The purpose of this section is not to describe in detail but to outline how a revenue share model operates and to describe the operational, commercial and market opportunities of this model with the deployment of MMS (Sadeh, 2002; Hilavuo, 2002). Let's use a practical example to outline the revenue share business model for MMS at the time of writing. Vodafone (UK) offer a revenue split of 60% for the content provider, to cover costs such as content generation, and 40% for the operator, to cover distribution and marketing costs. This compares with the i-mode service launched by KPN in The Netherlands and Germany where 86% is for the content provider and 14% for the operator.

BT's "click and buy" service is an example of a payment scheme that encompasses the mobile domain and provides a purchasing service that is independent of any network access. The revenue split is usually on a per-message basis. There will likely be further adjustments based on volume of traffic generated by a particular service. The pricing structure for premium MMS content will probably be in the region of 50p to £5.

Marketing demands (and accordingly costs) of promoting any multimedia messaging-based service should not be underestimated. Service usage is very sensitive to the amount of marketing effort in terms of advertising or promotional "free" usage periods. Often, short peaks in demand are seen after advertising a particular service. This demonstrates the competition that exists from alter-

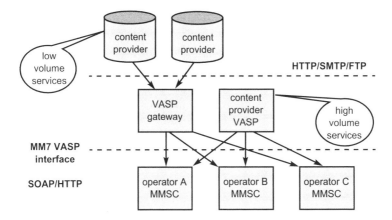

Figure 2.9 Outline of the VASP interface

HTTP = Hyper Text Transfer Protocol; SMTP = Simple Mail Transfer Protocol; FTP = File Transfer Protocol; VASP = Value Added Services Provider; SOAP = Simple Object Access Protocol; MMSC = Multimedia Messaging Services Centre

native services to meet customer demand, or that a service was poorly received as it fell short of customer expectation. This final point makes it difficult for a content provider to build a robust business case around a specific service.

Many mobile operators will offer an interface to a Value Added Services Provider (VASP) platform (Figure 2.9), rather than developing applications. The operator will be positioned as an aggregation and distribution mechanism to deliver third-party applications to the market. The MMS VASP API (MM7) will allow integration of applications from MMS content partners. A VASP will leverage the mobile operator's existing customer base and comparative advantage of lower customer acquisition and retention costs, and capitalize on its key enabling areas of message delivery, billing, CRM, customer registry and in the future presence and location information. The commercial opportunity for a VASP in revenue sharing with a mobile operator will lie in access to an installed customer base and the ability to charge customers a price per message that is higher than standard peer-to-peer SMS messages and even higher for premium SMS services (e.g., information and interactive services: see Stegavik, 2002).

The market opportunity for a VASP will be dictated by the rapid launch of new services: MMS devices are available and require the content and applications to persuade consumers to upgrade. Consumer demand is growing: today's lifestyles require consumers to be able to "do" more while mobile, access richer content than SMS text-only services and communicate via pictures (e.g., MobileYouth reported in October 2002 that 1.5 million MMs had been sent in Italy since the launch of the service earlier that year (**www.mobileyouth.org**).

One area mobile operators are very unlikely to enter on a revenue share basis is the standard data or airtime charges generated. Premium rate phone numbers are a possible solution to charging for voice services, and currently revenue sharing the premium cost of an MMS message is the only way to access operator revenues from MMS services. The introduction of MVNO services such as Virgin mobile and BT consumer mobile, using the T-Mobile network in the UK, are the primary routes to owning the customer relationship, billing and customer service, by contracting wholesale air and data access from a licensed mobile operator. The MVNO model has on the whole not been as successful in Asia: Virgin Mobile withdrew from Singapore in 2002 and the model is developing slowly in North America, notably Virgin Mobile (**www.virginmobileusa.com**), Boost Wireless (**www.boostmobile.com**) and Visage Mobile (**www.visagemobile.com**) began MVNO services in 2001. Interestingly, Visage Mobile has a business model based around creating MVNO services for companies in North America – an MVNO-enabler – rather than offering MVNO services directly itself.

2.5.2 Sponsorship business models

The use of advertising in MMSs will become commonplace, and the business model for delivering these will likely follow the fee-based, traffic-based or performance-based fee models of the fixed Internet. There are a number of issues and opportunities created by advertising that will be outlined in this section.

Where users have opted in to receive adverts based on personal preferences and these are sophisticated enough to use location within the context, there are tremendous opportunities. The example scenarios below show this. A user may have a personal preference for restaurant set to Italian and just before lunchtime when near a Pizza Hut (if occupancy is low on that day), they may be alerted to a discount on the lunch buffet. A user may be shown a product advertisement, delivered by MMS (Figure 2.10), in

Figure 2.10 MMS advertising (actual screen shot)
Source: **www.add2phone.com**

advance of being allowed to download a movie trailer video preview. The advertising fees paid by the advertiser to the operator may sustain a business model whereby the user can view the movie trailer free of charge.

The issues surrounding advertising mainly stem from unsolicited spam messages. In the UK the Independent Committee for the Supervision of Standards of Telephone Information Services (ICSTIS) has seen a 500% increase in 2002 of complaints about SMS marketing, most to do with premium number companies. In August 2002 it fined UK-based promotion firm Mobymonkey £50,000 ($80,000) for sending unsolicited text messages to users. The message said users had won a mystery prize and urged them to call a £1.50-per-minute number.

It is possible to configure some phones not to receive adverts: each MM has a message class (e.g., personal, advertisement, information and so on) and the default setting is personal. So, the receiving MMS client can check the message class to determine whether it is an advert and receive or not accordingly. These attributes could be abused by unscrupulous content providers, and there needs to be enforcement of this message class setting to prevent unsolicited adverts that do not conform. Currently, there are no implementations of audit and policy enforcement tools, but this is expected to change as the volumes of MMS traffic increase. Also, in theory it is possible for spam MMS to reach the scale encountered with fixed Internet email, although the operator controls many more of the interfaces for sending, receiving and billing MMS. It is still possible for a spammer to send an MMS to a user through a third-party MMSC outside the operator domain, and users (depending on the phone configuration) will automatically download the content at their expense.

2.5.3 Subscription business models

Although subscription models are seen by many users as a barrier to the take-up of some Internet services, the model has proliferated into MMSs. There are a number of examples where either using a subscription or per-use charging (and revenue share) business model is valid. Vodafone (Sweden) have partnered with many subscription-based content providers, including Reuters to deliver news information at 20 SEK for a 30-day period and ManagerZone offering games downloads at 10 SEK for 30-day access. O$_2$ (UK) offer Java games downloads on a per-game basis at a cost between £1.50 and £2.50 through its Revolution service (**www.o2.co.uk/ revolution**).

The addition of network elements to provide subscription management is essential, in order to enrol or delist a user once access has expired or been ceased. There are a number of suppliers of this infrastructure, including BrainStorm (UK), and clearinghouses that manage outsourcing of subscription management, such as Wmode.

2.5.4 Digital Rights Management (DRM)

DRM technology provides the ability to extract full commercial value from the content delivered by MMS. With current SMS messaging or WAP-based browsing, this value has been seen as limited, as the content delivered to the mobile handset was largely low value and generally free from copyright. With multimedia messaging, it is possible to enable a wider range of scenarios where convenience and immediacy mean the user sees value in paying an additional charge.

The sale of ringtones and icons to mobile phone users was estimated by NOPWorld (UK) in conjunction with Mobile Metrix (Sweden): the UK ringtone market they calculate will be worth over £60 million in 2003 (**www.mobile-metrix.com/press260203.html**). But as far as selling and delivering content via mobile phones is concerned, it is only the beginning. MMS could create a much larger and more diverse market both for message content and, more importantly, services delivered via MMS.

Although sold through the mobile operator, increasingly third-party content providers will be responsible for developing new services. So, operators' MMS systems must include facilities to manage payments to the content owners each time their content is used. They must also support the facility for users to forward on content to other users – for example, if a user receives and enjoys a cartoon, they might want to send it on to some of their friends. This is desirable for the operator, since it increases traffic, and for the content owner, since it increases exposure of its content. However, this so-called "superdistribution" of content also poses the danger that the owner will not be paid by many of the people receiving its content, just as the music industry recently experienced in the case of the peer-to-peer Internet distribution service Napster.

Mobile operators for the foreseeable future will be the primary distributors of content via mobile phones. They have the relationship with the end-customer and the infrastructure and systems in place required for high-volume billing. However, to retain that position, operators must be able to manage the digital rights of their content owners. DRM has so far been focused mostly on Internet solutions. With the advent of MMS, vendors are now beginning to develop mobile-specific DRM solutions (albeit still proprietary). Both equipment vendors such as NEC and Nokia (which licenses InterTrust products) and software companies such as Beep Science, Openwave and PacketVideo. These solutions are aimed at the specific sector that encompasses content policy Management systems, MMS client DRM support and video streaming DRM support.

Most DRM solutions are based around encrypting the content and issuing a voucher (often referred to as a mobile rights voucher) that decrypts the content according to the receiver's access rights, as shown in Figure 2.11 (Ikola, 2002).

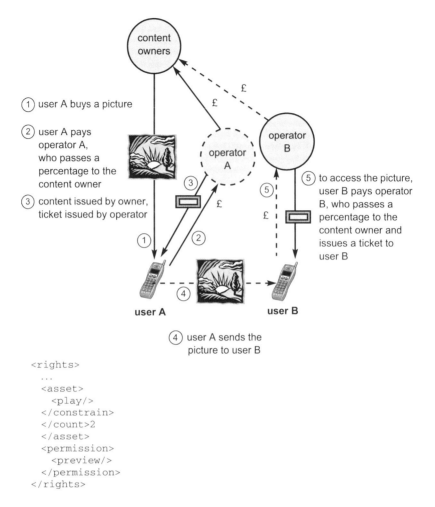

Figure 2.11 An example scenario of superdistribution
Source: Nokia website

A variety of different voucher types are needed to cope with the marketing requirements of content providers, such as:

- preview;

- outright purchase;

- one-time use (full or partial);

- multiple-time use (with discount support);

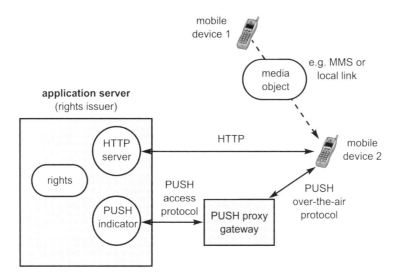

Figure 2.12 An example architecture for superdistribution

MMS = Multimedia Messaging Service; HTTP = Hyper Text Transfer Protocol

- limited validity period;

- adult-only access.

This voucher-based system enables control over superdistribution, since users receiving content from other users cannot access it without the appropriate voucher, which can only be obtained from the operator. As well as policing the owner's digital rights, this also has the advantage of establishing an interaction between the new recipient and the operators. So, with a DRM system in place, an operator can in effect enroll its users as a new channel of content distribution. The content protection market remains fragmented both for fixed and mobile phones. A standard for mobile DRM is essential for operators and content providers. At the time of writing, a large amount of work is being undertaken within the industry associated with DRM: OMA released a v1.0 DRM specification in July 2002 and Nokia have early implementations available to download in their mobile Internet development kit. The Nokia 6220 is the first model to support the OMA DRM specification (this is covered further in Chapter 7). Operators are, however, launching MMS – and enhanced services – with little or

no protection of content. Before DRM-supported services can reach
the mass market, a number of conditions must be met:

- handsets supporting DRM have to become widely available and
  they must have user interfaces that make access to protected
  content as easy as possible;

- superdistributed content should contain information about
  charges and usage rules;

- payments have to be collected with the help of widely available
  mechanisms and be transparent to the user.

It may be some time before all these requirements can be met.

2.6 CONCLUDING REMARKS

In this chapter, we took a look at the overall multimedia messaging
value chain and its many players, from infrastructure providers
and handset manufacturers to content providers and mobile operators. As we have seen, the relationships between these can be
complex. The primary business model for MMS is the sharing of
revenue on a per-transaction basis, although secondary business
models are still valid, such as advertising and subscription
billing, to name a few. Ultimately, the need for DRM will be
necessary to ensure a sustainable business model for content creation and distribution. This may prove to be the successful model to
drive fixed Internet services in the future.

Part II

The Technologies of Multimedia Messaging

3

A Standards-based Approach

3.1 INTRODUCTION

In this chapter we will describe the basic technical specifications for Multimedia Messaging Services (MMSs) and their implementation in mobile operator networks. There are some functional differences between the main vendors' MMS network products, but all of them conform to Third Generation Partnership Project (3GPP) MMS specifications. These are being developed by the 3GPP, with the main MMS platform vendors all active in the bodies carrying out the standardization work. Although it is a Third Generation (3G) standard, network operators across Europe and Japan are deploying MMS over General Packet Radio Service (GPRS) (or other packet-based) networks, using Wireless Application Protocol (WAP) as the transport.

The concept of a "partnership project" was pioneered by the European Telecommunications Standards Institute (ETSI) early in 1998 with the proposal to create a 3GPP that focused on Global System for Mobile communications (GSM) technology.

Although discussions did take place between ETSI and the ANSI-41 (CDMA/TDMA [Code Division Multiple Access/Time Dimension Multiple Access]) community with a view to consolidating collaboration efforts for all International Telecommunications Union (ITU) "family members", in the end it was deemed

Multimedia Messaging Service   Daniel Ralph and Paul Graham
© 2004 John Wiley & Sons, Ltd   ISBN: 0-470-86116-9-X

appropriate that a parallel partnership project be established, 3GPP2, which, like its sister project 3GPP, would embody the benefits of a collaborative effort (timely delivery of output, speedy working methods), while at the same time benefiting from recognition as a specifications-developing body. The primary focus of 3GPP2 is the air or radio interface and the migration (of mainly) US networks to cdma2000; this is also the main difference between 3GPP and 3GPP2, where in 3GPP focus is given to the migration from GSM to WCDMA networks.

As discussed, the MMS phones available in 2002 allow users to exchange messages that include still pictures, animations and sounds. In future, richer formats such as video will be added and Hyper Text Transfer Protocol (HTTP) may be used as the transport. MMS is a non-real-time service, designed to provide a similar user experience to that of existing services such as Short Message Service (SMS), but has been extended to include multimedia elements.

The key documentation in the standard are the functional specifications produced by the 3GPP's Technical Specification Group TS23.140. At the time of writing the current version of Release 5 of the document is V5.5.0. The hierarchy of documentation is defined by stages with increasing levels of detail at each stage, as shown in Figure 3.1. The basic definition of MMS services is undertaken by the Stage 1 group TS22.140, which is further supported by specifications on charging, media formats and detailed architecture.

MMS has been designed not to be specific to any network technology, and it can be implemented in a number of different technological frameworks. However, the WAP implementation is the most highly developed at present and most of the MMS products that are currently available incorporate it. The Open Mobile Alliance (OMA) (previously known as WAP Forum) has worked with the 3GPP on MMS standards from the outset, and its specifications are detailed in a series of documents (see Figure 3.1).

3.2 OVERVIEW OF MMS SPECIFICATIONS

The standard describes how service requirements are realized with selected technologies. As far as possible existing protocols (e.g., WAP, Simple Mail Transfer Protocol [SMTP], ESMTP for transfer

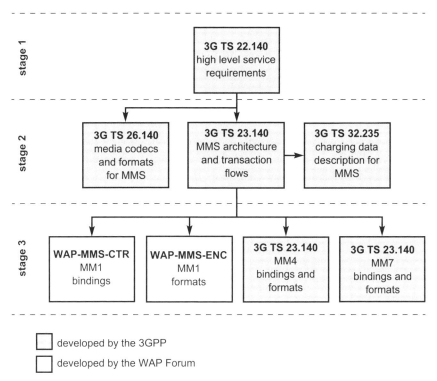

Figure 3.1 MMS technical specification overview

3GPP = Third Generation Partnership Project; WAP = Wireless Application Protocol;
MMSE = Multimedia Messaging Service Environment; VASP = Value Added Services Provider

and lower layers for push, pull and notification) and existing message formats (e.g., SMIL [Synchronized Multimedia Integration Language], MIME [Multipurpose Internet Email Extension]) are used for the technical realization of MMS.

3.2.1   To receive a message

A Multimedia Message (MM) is stored at the Multimedia Messaging Service Centre (MMSC) (similar to a Web server), which sends the recipient a notification message using WAP push (which is essentially a specially formatted binary SMS message). The notification message triggers the receiving terminal to retrieve the message automatically (or not, depending on settings defined by the user) by making a request to the configured WAP gateway, to retrieve the Uniform Resource Locator (URL) of the MMS content stored on the MMSC. This allows the receipt of the message to be transparent to the user, as is the case with SMS.

3.2.2   To send a message

Users typically create an MMS message on an MMS-enabled mobile phone and will enter the text and select the multimedia elements to be included with it. They can send the message to either a phone number or an email address. The message is sent to the user's MMSC by making a WAP Session Protocol (WSP)/HTTP POST from the MMS client through a WAP gateway. The MM can then either be sent to the receiving phone (via a WAP-PUSH/SMS notification message) or translated into an email and sent to the recipients email account, or posted to a website where the message can be viewed.

The content of MMS messages has been further clarified by the MMS Conformance Specification version 2.0.0 written by the MMS Interoperability Group (which currently consists of representatives from CMG, Comverse, Ericsson, Logica, Motorola, Nokia, Siemens and Sony Ericsson). The 3GPP MMS specification determines SMIL 2.0 Basic profile as the presentation format, an Extensible Mark-up Language (XML)-based standard that defines how the multimedia elements are coordinated. The suggested multimedia formats for 3GPP release 4 are:

- text;

- audio – Adaptive Multi Rate (AMR);

- image – JPEG, GIF 87a, GIF 89a, WBMP, BMP, PNG;

- Personal Information Manager (PIM) – vCard 2.1, vCalendar 1.0.

The following additional media formats are included in release 5 of the MMS standard:

- audio – MP3 and Midi (standards for hi-fi and professional quality sound);

- video – H.263 and MPEG4 (video compression standards pro-

	SMS	MMS
Message send protocol	SS7 and MAP	IP transport protocols, WAP encapsulation and SMTP/HTTP
Message delivery	One-stage delivery: message push	Two-stage delivery: notification SMS-based WAP-PUSH, followed by content retrieval using WSP/HTTP
Facilities for external applications	Requires multiple specific protocols and solutions: SMPP, CIMD2 and UCP	Uses standard Internet protocols: SMTP, HTTP and XML
Capacity planning	Load balancing is not easily applied to SS7; planning for heavy loads is based on platforms using closed, proprietary standards produced by specialist vendors	Scalability based on SMTP, which incorporates powerful load balancing and sharing features; widely available from multiple vendors
Media streaming	Not supported	Supported

Figure 3.2 Functional differences between SMS and MMS

SMS = Short Message Service; MMS = Multimedia Messaging Service; SS7 = Signalling System 7; MAP = Mobile Application Part; IP = Internet Protocol; WAP = Wireless Application Protocol; SMTP = Simple Mail Transfer Protocol; HTTP = Hyper Text Transfer Protocol; WSP = WAP Session Protocol; SMPP = Short Message Peer to Peer; CIMD = Computer Interface to Message Distribution; UCP = User Control Point; XML = Extensible Mark-up Language. Source: Comverse

duced by the ITU-T and International Organization for Standardization [ISO] respectively).

From the service point of view, MMS has been developed as the successor to SMS. It uses the same paradigm as SMS, pushing the whole message as a single package to the receiver's handset and storing it there. However, from the technical point of view, there are many important differences from SMS: the most fundamental being that, whereas SMS is carried in the out-of-band SS7 signalling infrastructure of Second Generation (2G) networks, MMS travels over the in-band network using IP transport (Figure 3.2). The standard defines three basic MMS message modes:

- Person-to-Person (P2P);

- Person-to-Machine (or service) (P2M);

- Machine (or service)-to-Person (M2P).

MMS is expected to play a leading role in the convergence of mobile networks and Internet-based email services. The billing functionality provided by the MMS specification is the best hope to continue the success of the billing model established by SMS.

   Initially, interoperability with email is important to MMS, the installed user base of handsets will need to grow over a two-year period within a new market in order to begin to mature. The ability to communicate to PC-based Internet email accounts will help spur this growth. The standard does not specify how MMS messages are exchanged with mobile phones that do not support MMS. There are two primary scenarios: delivering an MMS to an SMS-capable phone is done by sending a URL where the user can view the MMS in a Web browser; or an MMS can be delivered to a WAP-enabled phone, where the user can view text and still images in a WAP browser. The basic specifications used as building blocks for the development and deployment of MMS content are shown in Table 3.1.

3.3   3GPP RELEASE 5 – ARCHITECTURE OVERVIEW

3GPP has defined a reference architecture for MMS implementation, which specifies the functionality that vendors' MMS network products must support. This reference architecture, shown in Figure 3.5, contains a number of interfaces or reference points that describe interactions between the elements. Some of these are specified in release 5 of the specification, while others are not, though some may be specified in future releases, as detailed in the diagram.

3.4   WHAT IS AN MMSC?

The 3GPP's reference architecture for MMS forms the basis of the MMS network equipment being produced by vendors such as CMG, Comverse, Ericsson, LG Electronics, Logica, Motorola, Nokia and Siemens. In each case the core network element is the

Table 3.1 Outline of basic specifications and industry bodies

3GPP	www.3gpp.org/ftp/Specs/	TS 22-140: service aspects TS 23-140: functional specification TS 32.235: charging description for application services (release 4) TS 26.140: media formats and codecs TS 26.234: packet-switched streaming service protocols and codecs
W3C	www.w3.org/AudioVideo www.w3.org/TR/SVG www.w3.org/TR/SOAP	SMIL SVG SOAP
IETF	www.ietf.org/rfc/	RFC 2387: multipart/related MIME type RFC 2557: MIME encapsulation of aggregate documents RFC 2822: Internet message format RFC 821: SMTP
WAP Forum	www.openmobilealliance.org www.wapforum.org/what/technical.htm	WAP-MMS-ARCH: MMS architecture overview WAP-MMS-CTR: MMS client transactions WAP-MMS-ENC: MMS message encapsulation WAP-200-WDP: segmentation and reassembly (SAR) WAP-277-XHTMLNP: XHTML mobile profile WAP-248-UAPROF user agent profile WAP-182_ProvArch provisioning architecture

3GPP = Third Generation Partnership Project; SMIL = Synchronized Multimedia Integration Language; SVG = Scalable Vector Graphics; SOAP = Simple Object Access Protocol; MIME = Multipurpose Internet Email Extension; SMTP = Simple Mail Transfer Protocol; XHTML = Extensible Hyper Text Mark-up Language.

MMSC, which incorporates the relay, server and user database elements of the reference architecture. The MMSC is responsible for managing and monitoring MMS traffic, and for providing the data needed for charging and billing. The main functions of the MMSC are to:

- receive MMS messages and transmit them to their destinations;

- receive messages for MMS users from other services, such as email;

- convert the content of a message, where supported, to suit the capabilities of the destination device;

- generate the data records needed to bill subscribers for service usage and content.

The main components of the MMSC are:

- MMS relay – the hub of the system, this manages the flow of messages within the operator's network, between other operators' networks and between other networks such as the Internet (similar to an email system such as sendmail);

- MMS server – holds MMS messages until they can be delivered to their destination, similar to a web server like Apache (this is not permanent MMS storage although platforms that can offer this service are available, but they are not part of the MMSC specification);

- MMS user database – contains an operator's subscriber profile, with information such as the type of handset each subscriber is using, or authentication details.

3.5   BASIC FUNCTIONALITY OF AN MMSC

Each vendor's MMSC will be differentiated in its own way, but all MMSCs support a basic functionality that is necessary for MMS processing and interoperability. These functions of the MMSC are:

- Message notification – both message arrival and acknowledgement of message receipt.

- Message store and forward – the MMSC temporarily stores a message, while the receiver is notified of its arrival.

- Device capability – the MMSC interacts with the MMS user agent in order to determine the capabilities of the destination device. The most widely used mechanism for this is the WAP User Agent Profile (UAProf).

- Content transcoding – if the receiving device is not an MMS handset or is of a different type to the sender, the MMSC converts the media elements to a format that the receiving device can handle and, if necessary, delivers them as separate attachments or removes them accordingly.

- Address resolution – the MMSC needs to handle translation between phone numbers and email addresses.

- Multimedia streaming – for a video MM it will be necessary to stream media to the handset rather than download it; this could be used to support multicast and digital rights management. The implementation of MMS streaming is immature and discussed further in Chapter 4.

- Message billing data – the MMSC must provide the necessary information to construct billing data records and pass them to the operator's billing system. Event-based, media-based and volume-based charging models must be supported, as well as charging for third-party content and value-added service usage.

3.6 MMS ADDRESSING MODELS

The MMS architecture indicates one of the key advantages that MMS has over other forms of wireless messaging. The 3GPP specification stipulates that MMS must support sending messages to and from both mobile phone (MS-ISDN [Integrated Services Digital Network]) numbers and email addresses. In the network architecture MMS messages are passed between different operators networks as Internet Protocol (IP) traffic, using SMTP, so each operator's MMSC is assigned an Internet domain name (such as **mmsc.operator.com**). An MMS address provided by the user agent is either a routable email address (as specified in the Internet Engineering Task Force [IETF] RFC 822) or some other routable address such as an E.164 phone number (e.g., +447803232842, or

+447803232842/TYPE=PLMN@mmsc.vodafone.com, or an agreed alias like daniel.ralph@vodafone.com).

The mechanism through which MS-ISDN numbers are mapped to routable IP addresses is left up to the vendors and operators at present, although the 3GPP expects to specify it in future releases of the MMS standard, most likely using the ENUM global numbering proposal that is currently being developed by the IETF. The implementation of the addressing mechanism will have implications for Mobile Number Portability (MNP), which must be supported due to regulatory mandate; this permits users to move their MS-ISDN numbers to an alternative network.

MMS has included a specification to interwork with existing MNP intelligent network services (discussed in Chapter 5). Technical details of the MMS addressing model can be found in the WAP Forum's specification on MMS message encapsulation (see WAP-209).

3.6.1   Multimedia Messaging Service Environment

Figure 3.3 shows a generalized view of the MMS architecture. It combines different networks and network types, and integrates messaging systems that already exist within these networks. The terminal operates with the Multimedia Messaging Service Environment (MMSE), which may comprise Second Generation (2G) and Third Generation (3G) networks, 3G networks with islands of coverage within a 2G network and roamed networks. The MMSE provides all the necessary service elements (e.g., delivery, storage and notification functionality). These service elements may be located within one network or distributed across several networks or network types. Figure 3.4 shows that multimedia messaging may encompass many different network types. The basis of connectivity between these different networks is provided by the IP and its associated set of messaging protocols. This approach enables messaging in 2G and 3G wireless networks to be compatible with messaging systems found on the Internet.

3.6.1.1   MMSNA

The Multimedia Messaging Service Network Architecture (MMSNA) encompasses all the various elements that provide a

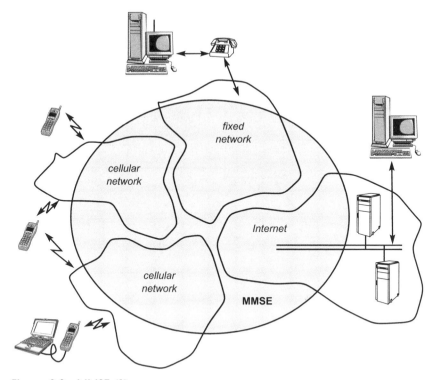

Figure 3.3 MMSE (1)

MMSE = Multimedia Messaging Service Environment. Source: 3GPP (2002a)

complete MMS to a user (including interworking between service providers). The MMSE is a collection of MMS-specific network elements under the control of a single administration. In the case of roaming the visited network is considered a part of that user's MMSE. However, subscribers to another service provider are considered to be a part of a separate MMSE.

3.6.1.2 MMS relay/server

The MMS relay/server is responsible for storage and handling of incoming and outgoing messages and for the transfer of messages between different messaging systems. Depending on the business model, the MMS relay/server may be a single logical element or may be separated into MMS relay and MMS server elements; these may be distributed across different domains.

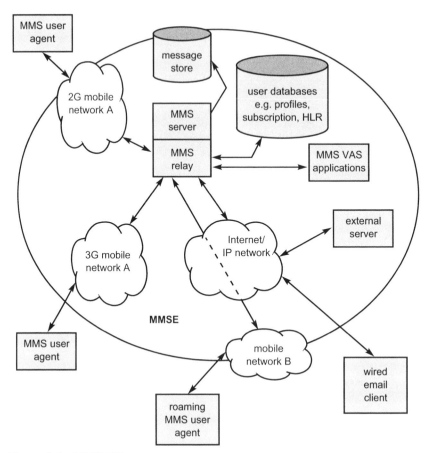

Figure 3.4 MMSE (2)
MMSE = Multimedia Messaging Service Environment; VAS = Value Added Service. Source: 3GPP (2002a)

3.6.1.3 MMS user databases

MMS user databases may be comprised of one or more entities that contain user-related information such as subscription and configuration (e.g., user profile, Home Location Register [HLR]).

3.6.1.4 MMS user agent

The MMS user agent resides on User Equipment (UE), a Mobile Station (MS) or on an external device connected to a UE/MS. It is an application layer function that provides users with the ability to view, compose and handle MMs (e.g., submitting, receiving or deleting them).

3.6.1.5 MMS VAS applications

MMS VAS applications offer value-added services to MMS users. There could be several MMS VAS applications included in or connected to an MMSE. Such applications may be able to generate Charge Detail Records (CDRs).

3.6.1.6 MMS reference architecture

Figure 3.5 shows the MMS reference architecture and identifies those reference points within an MMSNA that are described in the following list. The interfaces in the MMS reference architecture are:

MM1 The reference point between the MMS user agent and the MMS relay/server.

MM2 The reference point between the MMS relay and the MMS server.

MM3 The reference point between the MMS relay/server and external (legacy) messaging systems.

MM4 The reference point between the MMS relay/server and another MMS relay/server that is within another MMSE.

MM5 The reference point between the MMS relay/server and the HLR.

MM6 The reference point between the MMS relay/server and the MMS user databases.

MM7 The reference point between the MMS relay/server and MMS VAS applications.

MM8 The reference point between the MMS relay/server and a billing system.

It is useful to highlight just two of the key interfaces that we will return to later in the book: the MM4 interface that supports interoperability between MMSEs and the MM8 interface between the MMSE and the billing system that supports collection of CDRs. These are discussed further in Chapter 5.

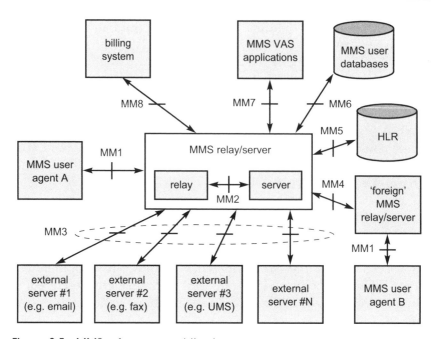

Figure 3.5 MMS reference architecture

MMS = Multimedia Messaging Service; VAS = Value Added Service; HLR = Home Location Register; UMS = Unified Messaging System

3.7 MM4: INTERWORKING OF DIFFERENT MMSEs

Reference point MM4 between MMS relay/servers that belong to different MMSEs is used to transfer messages between them. Interworking between MMS relay/servers is based on SMTP according to STD 10 (RFC 821) as depicted in Figure 3.6 and 3.7. The MMS relay/server should be able to generate charging data (CDR) when receiving MMs from or when delivering MMs to another element of the MMSNA according to 3GPP (2002b). It should also be able to generate charging data for Value Added Service Provider (VASP) related operations.

3.8 MM7: MMS RELAY/SERVER – MMS VAS APPLICATIONS

Reference point MM7 is used to transfer MMs from the MMSC to MMS VAS applications and to transfer MMs from MMS VAS

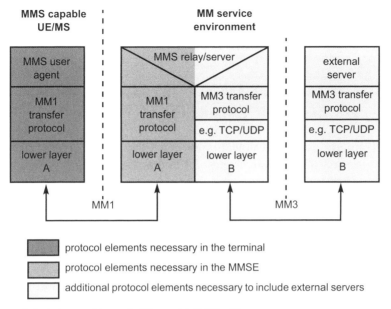

Figure 3.6 Interworking of different MMSEs (1)

MMSE = Multimedia Messaging Service Environment; UE = User Equipment; MS = Mobile Station; MM1 = interface to MMS relay/server; MM3 = interface to external messaging systems; TCP = Transmission Control Protocol; UDP = User Datagram Protocol. Source: 3GPP (2002a)

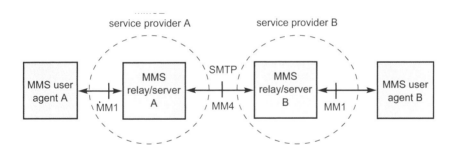

Figure 3.7 Interworking of different MMSEs (2)

MMSE = Multimedia Messaging Service Environment; MM1 = interface to MMS relay/server; MM4 = interface to another MM5 relay/server in another MMSE. Source: 3GPP (2002a)

applications to the MMSC. This reference point is based on existing protocols (e.g., SMTP or HTTP/SOAP [Simple Object Access Protocol]) for release 5 of the specification; future releases may propose a mandatory protocol and encoding schemes. The service provider may decide to use an encoding format in this reference point that uses the encoding implementation used in the MM1 reference point.

3.9 EXAMPLE OF MMS INTERACTION WITH 2G/3G VOICEMAILBOXES

MMS interaction with voicemailbox systems should be performed on a non-real-time basis. Figure 3.8 illustrates an example architecture for the incorporation of voicemailboxes. The Voice Profile for Internet Mail version 2 (VPIMv2) provides format extensions for MIME, supporting the transmission of voice messages over standard Internet email systems. The VPIM concept was developed by the Electronic Message Association (EMA), and after VPIMv2 had been reviewed by the IETF it became RFC 2421. The VPIM specification allows voice records to be MIME-encapsulated and sent as Internet mail attachments via SMTP or retrieved as Internet mail attachments via Post Office Protocol version 3 (POP3) or IMAP4. The MIME type used for voice messages is "audio/*". For the interaction of MMS with voicemailboxes, the voice mailbox may forward received voice records as VPIM messages via SMTP to the MMS relay/server, implying that voice message downloads are always done via the MMS. In this case the protocol

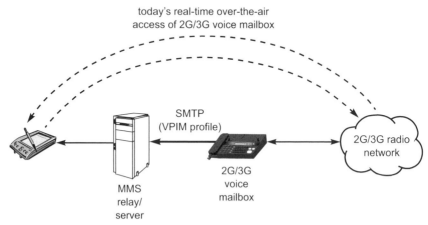

Figure 3.8 First example of MMS interaction with 2G/3G voicemailbox, based on VPIM

MMS = Multimedia Messaging Service; VPIM = Voice Profile for Internet Mail; SMTP = Simple Mail Transfer Protocol; 2G/3G = Second and Third Generation

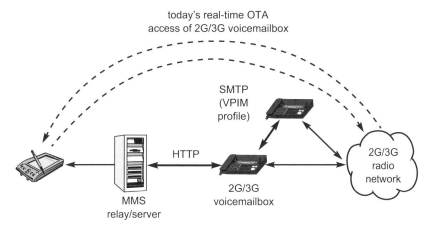

Figure 3.9 Second example of MMS interaction with 2G/3G voicemailbox, based on HTTP

MMS = Multimedia Messaging Service; HTTP = Hyper Text Transfer Protocol; 3G/4G = Second and Third Generation; SMTP = Simple Mail Transfer Protocol; VPIM = Voice Profile for Internet Mail

to be used on the interface between MMS relay/server and the voicemailbox is SMTP and thus is identical to the one used between different MMS relay/servers.

Alternatively, the MMS relay/server may poll the voicemailbox via POP3 or IMAP4 for new messages received. Messages the user wants to retrieve via the MMS can then be downloaded via POP3/IMAP4 from the voicemailbox to the MMS relay/server, from where they are delivered to the MMS user agent. This enables the user to do both: retrieve voice messages via today's real-time voicemail services or as MMs. In any case it is expected that the voicemailbox is still the owner of the message and as a consequence responsible for the storage. Another alternative could be MMS interworking with a 2G/3G voicemailbox system through an HTTP interface (as depicted in Figure 3.9).

3.10 OPEN MOBILE ALLIANCE – THE WAP STANDARD

The WAP standard is a result of continuous work to define an industry-wide specification for developing applications that operate over wireless communication networks. The scope for the OMA is to define a set of specifications to be used by service applications as a result of the wireless market growing very quickly and reaching new customers and services. To enable operators and manufacturers to meet the challenges in advanced services, differentiation and fast/flexible service creation, OMA defines a set of protocols in transport, security, transaction, session and application layers.

MMS is a system application by which a WAP client is able to provide a messaging operation with a variety of media types. The service is described in terms of actions taken by the MMS client through a WAP gateway to the MMS relay/server. The service description of the MMS can be found in "Multimedia Messaging Service v1.1" (OMA-MMS-V11). This specification defines the message encapsulation (i.e., the message structure and encodings for the MMS). The OMA specifications (de facto standards) are often referred to as the technical realization of the 3GPP specifications. Current mobile network implementations of the MMS infrastructure are intrinsically dependent on the WAP gateway and WAP Push Proxy Gateway (PPG) network elements to support the notification and transmission of MMs from a WAP-enabled client to the MMSC. For additional information on the WAP architecture, refer to "Wireless Application Protocol Architecture Specification" (**www.wapforum.org** or **www.openmobilealliance.org**).

3.10.1   Architectural support for MMS

WAP support for MMS is based on the services provided by its supporting technology. The scope of WAP, as far as MMS is concerned, is shown in Figure 3.10, but does not cover activities or network elements beyond those shown. The figure shows an MMS relay/server, which in WAP architecture terminology is referred to as an MMS server. The WAP architecture also refers to the MMS user agent as an MM client. These are equivalent functionalities. Figure 3.10 shows two links: the first, between the wireless MMS user agent and the WAP gateway, is where the "WAP stack" is used to provide a common set of services over a variety of wireless bearers. For application-oriented services, like MMS, the interest is primarily in services offered by WSP. The second link connects the WAP gateway and the MMS relay/server. In the WAP architecture the MMS relay/server is considered an origin server (or Web server/email system). These entities

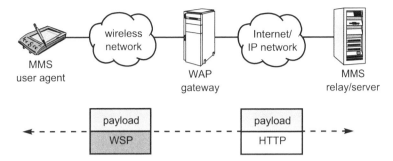

Figure 3.10 Scope of WAP support for MMS

WAP = Wireless Application Protocol; MMS = Multimedia Messaging Service; IP = Internet Protocol; WSP = WAP Session Protocol; HTTP = Hyper Text Transfer Protocol. Source: OMA MMS-v1.1

are connected over an IP network such as the Internet or a local intranet and HTTP is used for data transfer between the MMSC and the WAP gateway.

End-to-end connectivity, for delivery of an MM, from the wireless MMS user agent to the MMS relay/server is accomplished by sending data over WSP (to the WAP gateway), which then transmits HTTP (to the MMSC). This is accomplished by both using the WSP/HTTP POST method for sending messages, originating at the wireless MMS user agent, and by using the WAP push access protocol to receive a notification of a message in the other direction.

The WAP gateway, which enables the needed interworking, should not modify data transfer via these transactions. The WAP view of MMS is restricted to the interactions between the MMS user agent and the MMS relay/server. It makes no representations to services that are provided to or required of any other network elements.

3.11 MMS AND WAP-PUSH

The WAP-PUSH implementation (Figure 3.11) of MMS notification uses the PPG functionality that appears in WAP 2.0 and is also included in a more limited form in WAP 1.2. MMS uses WAP PUSH to control the user's experience of receiving a message: the most common implementation is that the user receives a notification of message arrival (via SMS), with an option (preconfigured in the MMS settings) to initiate retrieval of the message to the handset immediately, or to do so later on.

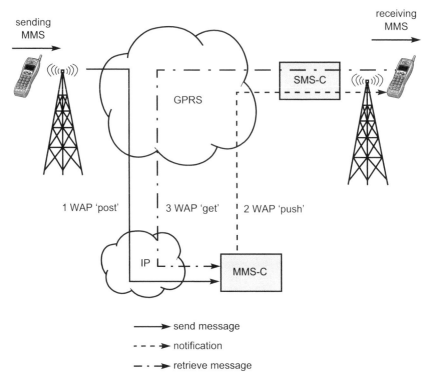

Figure 3.11   WAP-PUSH overview

WAP = Wireless Application Protocol; MMS = Multimedia Messaging Service; SMS = Short Message Service; GPRS = General Packet Radio Service. OMA MMS-v1.1

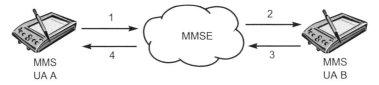

Figure 3.12   MMS call flow

MMSE = Multimedia Messaging Service Environment; UA = User Agent

It is useful at this stage to show the message flows (Figure 3.12) between a mobile device, the WAP gateway, the Short Message Service Centre (SMSC) and the MMSC; this will help support the description of what is a set of complex interactions involved in the submission and subsequent delivery of an MMS through the mobile network.

3.12 FUTURE DEVELOPMENTS

At the time of writing the current version of TS 23.140 release 5 (completed June 2002) identifies areas that require further elaboration. The following subsections describe the primary items planned to be included in release 6 of MMS.

3.12.1 MMS user agent (MM1)

- Further enhancements to MM1 include the addition of information elements on MM1 to improve acknowledgements, time stamp/time zone and adding data such as a message digest (e.g., displaying the text of an MM in the notification or adding user prompts in MMS);

- changes to MMbox handling (e.g., selective retrieval of MM objects and storage using a directory structure);

- a non-WAP MM1 implementation (for MM retrieval without the WAP infrastructure).

3.12.2 Interworking with legacy systems (MM3)

- MM3 interface to an external mail server to allow standardized interworking with legacy voicemail systems/voice messaging (voice XML [VXML]);

- a definition for legacy handset support of handling unsupported message formats (e.g., an MMS sent to a WAP-only phone).

3.12.3 Interconnection with different MMSEs (MM4)

- Further enhancements to MM4 to facilitate discovery of external MMS relay/server capabilities.

3.12.4 User database-related items (MM6)

- Currently, this remains undefined and further progress is to be made on the specification for MM6 transactions.

3.12.5   Value Added Service Provider-related items (MM7)

- Enhancements to MM7 (e.g., defining APIs, handling mailing (distribution) lists, better handling of financial transactions, improved authorization/authentication);

- investigate and identify support for enhancements of the interworking with VAS applications (e.g., based on OSA [Open Services Architecture] extensions);

- VASP interworking (e.g., MMSE_A offering services to subscribers of MMSE_B).

3.12.6   Billing-related items (MM8)

- Define MM8 reference point;

- prepaid interface for MMS (in particular, to prevent transmission if the user has a low account balance);

- charging enhancements (third party pays, prepaid charging, etc.);

- advice of charge to user;

- support for MM volume classes-based charging.

3.12.7   End-to-end service items

- support for Digital Rights Management (DRM) in MMS;

- support for end-to-end security and privacy (e.g., Virtual Private Network [VPN]/IPSECurity – an encryption protocol specified by the IETF – terminal security);

- support for filters (e.g., spam control);

- 3GPP Generic User Profile (GUP) interworking, for holding MMS UA capabilities, offering two-way capability detection/negotiation and support for notification to legacy terminals if an MMS-user changes to a non-MMS terminal;

- interface to a presence server and a presence-enabled (online) address book;

- investigate and identify relationship between MMS and instant messaging;

- investigate possibility of transcoding interface to external transcoding server;

- define MMS Over-The-Air (OTA) provisioning.

These lists represent a fairly lengthy plan of activity to be undertaken in 2003/2004 to further develop MMS standards. Obviously, some items may be dropped and others introduced, depending on the commercial importance of such a change. Change requests to release 5 generated by interoperability testing will be required to clarify aspects of the current version of the standard.

4

Application Layer

4.1 INTRODUCTION

In this chapter we will explore application capabilities and limitations, from devices and development environments to deployment
and testing. As there is a great deal of reuse of existing Internet
Protocols (IPs) and standards in the development of Multimedia
Messaging Service (MMS) applications, this will help an experienced developer to become comfortable with the syntax and
methodology used in the development phase. We will highlight
those areas that may be unfamiliar and explore the new techniques
introduced to tackle some of the deployment requirements.

   The chapter then continues with an outline of the terminals available in the market and the device requirements to deliver MMS.
Afterwards we discuss the User Agent Profile (UAProf) and capability negotiation that are used to determine device characteristics.
The issues are then outlined in the section on device presentation
and the potential solution by the use of content adaptation (or
transcoding). We then provide an overview of the Synchronized
Multimedia Integration Language (SMIL) development mark-up
language used to specify the content delivered in a Multimedia
Message (MM). There follows a section on the application development environments at Nokia and Ericsson. The next section is
on application deployment, which includes a range of topics,
from provisioning devices and testing applications to a discussion
of the importance of the MM7 interface in deploying MMS

Multimedia Messaging Service   Daniel Ralph and Paul Graham
© 2004 John Wiley & Sons, Ltd   ISBN: 0-470-86116-9

applications. The chapter is brought to completion with an overview of how MMS streaming of video content is enabled using SMIL.

4.2 WHAT IS SCALABLE VECTOR GRAPHICS?

The W3C (World Wide Web Consortium) Scalable Vector Graphics (SVG) 1.0 specification defines an Extensible Mark-up Language (XML) grammar for describing resolution-independent, two-dimensional graphics on the Web. The concepts in SVG are well suited to the application areas that are typical of mobile devices, including location-based services (e.g., displaying a map and route to the nearest parking station), infield Computer Aided Design (CAD)/engineering diagrams (e.g., a diagram of a circuit board with hyperlinked component descriptions) and entertainment (e.g., cartoon-like animation). However, the SVG 1.0 specification is primarily focused at desktop machines and as such does not specifically address the needs of devices with resource limitations such as mobile devices. For example, a typical Personal Digital Assistant (PDA) has a palm-sized display with low resolution, and a mobile phone has potentially extreme limitations on power and memory consumption. Both may be limited to an intermittent low-bandwidth network connection. These restrictions make it difficult for such devices to render the full range of SVG 1.0 content.

There are a number of MMS browsers that support SVG tags (Figure 4.1) for the display of graphics (including **www.annyway.com** and **www.bitflash.com**). It must be said that SVG is unsuitable for the display of photographic-type images.

4.3 WHAT IS MIME?

Although Multipurpose Internet Email Extension's (MIME) codings were originally developed for email in 1992, the same scheme was also adopted for browser use. The Internet Engineering Task Force's (IETF) RFC-2045 and related documents contain detailed information.

Servers identify the type of multimedia data they're sending by telling the browser the MIME type: short codes that say, "this is a PNG, this is an MP3, this is Shockwave" and so on. The server tells the browser (or other receiving application) which type of data

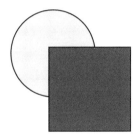

```
<svg width="128" height="128">

  <circle cx="45" cy="45" re"40" fill="grey" />
  <rect x="45" y="45" width="90" height="40" fill="black" />
  <text x="5" y="120">Scalable Vector Graphics</text>

</svg>
```

Figure 4.1 SVG example

format it is by putting MIME identifiers in the data it sends. When browsers render this content, they need to know what type they are rendering so that they can then match this media type against its list of installed plug-ins, to see which plug-in can play this type of content. If the browser has to guess, it might fail. An example of a MIME content-type (stored in the MIME header) for MMS is:

```
Content-type: application/smil
```

The mobile device would then use the MMS client to display MIME content of this type. It should be noted that the Multimedia Messaging Service Centre (MMSC) must be configured to allow retrieval of each different MIME type allowed.

MIME also uses other parameters in the header component that relate to the enclosed message and are often used in processing the message at the receiving application. An example of one of many MIME header parameters is:

```
X-MMS-Message-type: m-retrieve-conf
```

The MIME header is encoded in binary format and attached to the enclosed content for transmission. Further clarification of the use of content-type in both the MMS header and the Hyper Text Transfer Protocol (HTTP) header is shown in Figure 4.2.

If you were to store an MMS-encoded file, called testfile.mms, on a Web server for retrieval from an MMS phone, then the Web

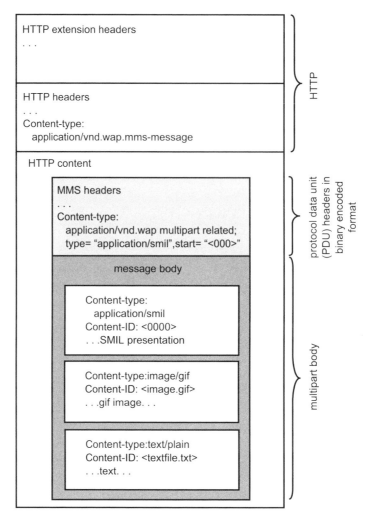

Figure 4.2 MMS encapsulation as stored on the MMSC

MMS = Multimedia Messaging Service; MMSC = MMS Centre; HTTP = Hyper Text Transfer Protocol

server would need to be configured to set the content-type in the HTTP header by editing the MIME-type configuration file to include the associated extensions:

Associated extension: Content type (MIME):

```
.wbmp               image/vnd.wap.wbmp
.wml                text/vnd.wap.wml
.mms                application/vnd.wap.mms-message
```

So, when a Wireless Application Protocol (WAP) gateway retrieves a file with the extension .mms it will receive an HTTP response header with:

```
Content-type: application/vnd.wap.mms-message
```

The MMS client will receive the MM (the HTTP header will have been stripped off at the WAP gateway) that will have:

```
Content-type: application/smil
```

4.4 WHAT IS XHTML MOBILE PROFILE?

Essentially, XHTML (Extensible Hyper Text Mark-up Language) is the mark-up language used to build pages displayed in the WAP browser of a mobile device. Its potential, however, is very much more. XHTML, according to the W3C, is the first major change to HTML since HTML 4.0 was released in 1997. The latest version of HTML (version 4.1) forms the basis for XHTML: all tag definitions and syntax are the same. XHTML adds modularity and enforces strict adherence to language rules and, as a result, XHTML brings a strict structure to web pages, which is especially important given the small screens and limited power of mobile devices. W3C is recommending XHTML for all future Web development for desktops as well as all other devices, including mobile handsets. The mark-up language's family tree:

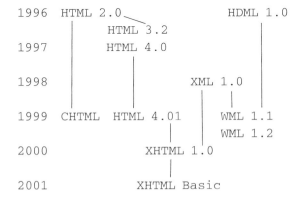

HDML = Hand-held Device Mark-up Language; CHTML = Compact HTML; WML = Wireless Mark-up Language; XHTML = Extensible HTML

XHTML Basic is the mobile version of XHTML 1.0 and is designed for Web clients that do not support the full set of XHTML features (e.g., Web clients such as mobile phones, Personal Digital Assistants (PDAs), pagers and set-top boxes). With XHTML Basic, a document can be presented on the maximum number of Web clients, including a wide range of mobile phones featuring different display formats and presentation capabilities.

XHTML Mobile Profile is defined by WAP Forum: it starts with XHTML Basic and adds some elements and attributes from full XHTML 1.0 that are useful in mobile browsers, including additional presentation elements and support for internal style sheets. XHTML Mobile Profile is a strict subset of XHTML and is the mark-up language of WAP 2.0.

4.5 WHAT IS SOAP?

Simple Object Access Protocol (SOAP) is an Extensible Mark-up Language (XML)-based messaging protocol. It defines a set of rules for structuring messages that can be used for simple one-way messaging, but is particularly useful for performing RPC-style (Remote Procedure Call) request–response dialogues. It is not bound to any particular transport protocol though HTTP/TCP (Transmission Control Protocol) is the most common implementation. Its most significant benefit is that it is not tied to any particular Operating System (OS) or programming language; so, theoretically, the clients and servers can be running on any OS or platform and be written in any language as long as they can encode and understand SOAP messages.

As such it is an important building block for developing distributed applications that exploit functionality published as services over an intranet or the Internet.

A SOAP developer's approach to the problem outlined in Figure 4.3 is to encapsulate the database request logic for the service in a method (or function) in C++, Visual Basic or Java, then set up a listener process that waits for requests to the service; such requests arrive in the SOAP format and contain the service name and any required parameters. As mentioned, the transport layer is normally HTTP; however, it could be implemented using Simple Mail Transfer Protocol (SMTP) or another transport layer protocol. Now, the listener process decodes the incoming SOAP request

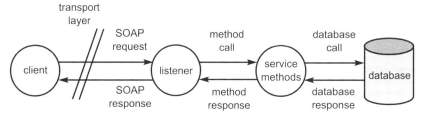

Figure 4.3 SOAP basis architecture
SOAP = Simple Object Access Protocol

and transforms it into an invocation of the method. It then takes the result of the method call, encodes it into a SOAP message (response) and sends it back to the requester.

We will now outline a more complex example of using SOAP to interact between distributed objects using different platforms and programming languages. In Figure 4.4 the developer has written the service method in Java and has connected to the database using an Oracle implementation of JDBC (Java Data Base Connectivity). The listener process is a Java servlet running within a servlet engine such as Tomcat. The servlet has access to some Java classes capable of decoding and encoding SOAP messages (such as Apache SOAP for Java) and is listening for those messages as a POST method from an HTTP/TCP request. The client is an Excel spreadsheet. It uses a visual basic macro which in turn exploits the Microsoft SOAP toolkit to encode a SOAP request and decode the response received (Soapuser, 2002).

4.6 MESSAGING TERMINALS

In this section we will explore the range of different terminal types available that support the MMS standard; these include smartphones, PDA devices and gaming devices. The main components that dictate the usability of a messaging terminal are:

- screen size
    - 160×120 (many phones offer this resolution)
    - 320×240 (many PDAs offer this resolution, referred to as QVGA [Quarter VGA]);
- battery life;

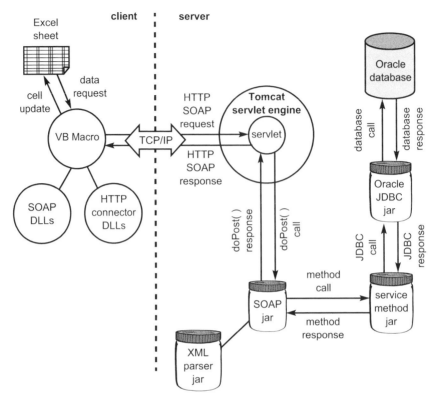

Figure 4.4 SOAP detailed architecture

SOAP = Simple Object Access Protocol; VB = Visual Basic; TCP = Transmission Control Protocol; IP = Internet Protocol; HTTP = Hyper Text Transfer Protocol; DLL = Dynamic Linked Library; JDBC = Java Data Base Connectivity; XML = Extensible Mark-up Language

- input method

  ○ pen
  ○ QWERTY keypad
  ○ number keypad;

- software capability

  ○ MMS client
  ○ WAP browser
  ○ WWW browser
  ○ SSL (Secure Sockets Layer) support
  ○ Symbian OS 7.0
  ○ video player
  ○ J2ME (Java 2 Micro Edition);

- network capability
  - ○ General Packet Radio Service (GPRS) $(4+2, 3+1, 4+4)$
  - ○ Third Generation (3G) (64–384kbps)
  - ○ Wireless Local Area Network (WLAN) (2–11Mbps).

These different permutations of presentation, input and access all offer interoperability challenges and are discussed further in this chapter.

4.6.1 Pogo nVoy e100

The nVoy e100 device (Figure 4.5) is the mobile part of the nVoy communicator solution. It features a large QVGA (320 × 240) colour TFT display with touch screen, integrated camera and Global System for Mobile communications (GSM)/GPRS mobile communications, all within a small, slim, elegant device that fits in the palm. The large, colour touch screen enables convenient viewing of Short Message Service (SMS), email and MMS messages, Internet pages, photos and allows the device to be used as a full colour digital camera. The touch screen contributes to the simplicity of use by allowing the user to directly click on objects and links on the screen.

Figure 4.5   Pogo-tech.com nVoy e100
Source: Pogo.tech, **www.pogo-tech.com**

4.6.2 Nokia N-Gage

Nokia N-Gage is a mobile game deck with communication and entertainment features. The game deck is fitted with an ARM processor, expansive memory, colour display and unique key-

Figure 4.6   Nokia N-Gage
Source: Nokia, **www.nokia.com**

board. The Multimedia Memory Card (MMC) memory card
support allows expanded software capabilities, and Digital Rights
Management (DRM) helps protect content. The device also builds
on the open Symbian 60 Platform and supports advanced mobile
business and lifestyle applications. As a communication device,
Nokia N-Gage game deck offers triband (GSM/EGSM) capabilities,
SMS and MMS messaging and full email support (IMAP4, POP3,
SMTP and MIME2). Users have access to mobile content with the
XHTML Mobile Profile and WAP 1.2.1 browser. For local connec-
tions, Nokia N-Gage game deck provides both Bluetooth and USB
(Universal Service Bus) capabilities. Network support is HSCSD
(High Speed Circuit Switched Data) and GPRS (2 + 2, 3 + 1, class
B and C.). The game deck features audio capabilities, including a
stereo FM receiver, downloadable polyphonic MIDI (Musical In-
strument Digital Interface) ring tones and a digital music player
(MP3 and AAC) with 64MB of storage (on an optional MMC).
The screen is 176×208 pixels, with 12-bit colour depth or 4,096
colours. Developers can choose to work in Java MIDP (Mobile
Information Device Protocol) or in C++ for the underlying
Symbian OS.

4.6.3   Ericsson P800

The P800 features a 208×320 colour touch screen and an integrated
digital camera. It occupies a form factor similar to that of a tradi-

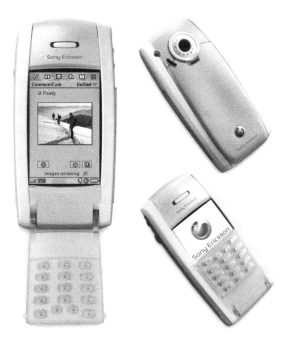

Figure 4.7   Ericsson P800
Source: Sony–Ericsson website

tional mobile telephone, using a flip cover to protect the screen and house the dialling buttons. This phone runs the Symbian OS 7.0 with 12MB of RAM, a built-in memory stick slot and a full Web browser. It has triband GPRS communications capabilities and includes Bluetooth. Symbian OS 7.0 provides support for mobile Java applications, MMS, Web browsing, WAP and i-mode. It includes a suite of PIM (Personal Information Manager) applications, viewing software for popular office documents and can store up to 200 digital images. With the P800 it is possible to take digital pictures, view them on the 208×320-pixel colour screen, store them in the photo album and send them as an email to a PC or as an MMS message to another phone. The P800 can also show a colour picture of the person who is calling.

The P800 offers the possibility of downloading and viewing video clips such as a sequence from a sports event, music video or movie trailer. It is also an organizer, handling daily operations such as calendar, email, address book and to-do lists. All these features can easily be synchronized with the most common office applications on a PC. The P800 enables note-taking and viewing files such as PowerPoint, Word or Excel. Included with Bluetooth wireless technology, the P800 can be connected wirelessly to a Bluetooth headset, a PC or other Bluetooth-enabled gadgets. The P800 incorporates the UIQ pen-based user interface and supports the downloading of applications, such as games, based on Java and C++. This opens up possibilities in several applications areas, and it means that consumers will be able to update their handsets regularly with new applications and content.

4.6.4   Device and client requirements

The MM1 interface supports communication between an MMS client and the MMSC (through a WAP gateway). It is not intended to outline the protocol used for MM1, although discussion of implementation specifics, such as notification and retrieval, are in the section on applications development. The interface further supports authentication of the terminal and terminal capability negotiation. We will discuss these areas further in this section.

4.6.4.1   MM1: MMS relay/server and MMS user agent

To provide implementation flexibility, integration of existing and new services together with interoperability across different networks and terminals, the MMS user agent uses the MM1 protocol framework to communicate with the MMS relay/server. In this framework the MMS user agent communicates with the MMS relay/server, which may communicate with external servers. The MMS relay/server provides convergent functionality between external servers and other MMS user agents and thus enables the integration of different server types across different networks.

The reference point MM1 is used to define the communication protocol between the MMS client (or user agent) and the MMSC in broadly the following activities:

- submit MMs from the MMS user agent to an MMS relay/server;

- retrieve (or pull) MMs using the MMS user agent from the MMS relay/server;

- submit push information from the MMS relay/server about MMs to the MMS user agent as part of an MM notification;

- exchange delivery reports between MMSC and MMS user agents.

Details for implementation of the MM1 transfer protocol using WAP or applications conforming to Mobile Station Application Execution Environment (MExE) (e.g., Java and TCP/IP) are elaborated within TS 23.140 (3GPP, 2002a). The WAP implementation option is described in WAP-205. Implementations based on applications using MExE may be defined in detail in future releases. Other implementations (e.g., using other standardized Internet protocols, such as HTTP) are not defined in the current version of TS 23.140 release 5 (3GPP, 2002a).

4.6.4.2 MMS user agent operations

The MMS user agent will provide, as a minimum, the following application layer functionalities:

- MM retrieval (MM delivery to the MMS user agent);

- terminal capability negotiation.

There are few phones in the market that only support this minimum feature set, the Nokia 3510 being one. Many vendor implementations of the MMS user agent provide additional application layer functionalities, such as:

- MM creation and composition;

- MM submission;

- MM preview;

- notify user of MM size prior to submission;

- user notification of arrival of an MM;

- digital signing of an MM on an end-user to end-user basis;

- decryption and encryption of an MM on an end-user to end-user basis;

- management and presentation of MMBox content;

- manipulation and storage of MMs on the terminal;

- configuration of MMS-related information on the (U)SIM;

- configuration of external devices;

- user profile management.

This optional list of additional functionalities of the MMS user agent is not exhaustive.

MM1 is an important interface and differs from the existing Internet protocols that support email delivery, such as Post Office Protocol version 3 (POP3) and SMTP. It is better suited to the mobile environment and its additional responsibilities include:

- personalizing the MMS, based on user profile;

- media type and format conversion;

- checking terminal availability;

- address translation;

- support of prepaid customers (e.g., advice of charge, notification of low balance, controlling reply-charging);

- control of multimedia streaming sessions.

4.6.4.3   Device capability profile structure

A device capability profile is a description of the capabilities of the device and can also be the preferences of the user of that device. It can be used to guide the adaptation of content presented to the device. A device capability profile for a mobile device is stored in a Resource Description Framework (RDF) document that follows the structure of the Composite Capabilities/Personal Preferences (CC/PP) framework and is queried by a CC/PP application; in the case of MMS this is specified by the WAP Forum and called UAProf (or User Agent Profile).

It is important to briefly describe the term CC/PP. Attributes are used for specifying device capabilities and user preferences. A set of attribute names, permissible values and semantics constitute a CC/PP vocabulary. An RDF schema defines a vocabulary. The syntax of the attributes is defined in the schema but so is to some extent the semantics. A profile is an instance of a schema and

```
[ex:MyProfile]
  |
  +--ccpp:component---> [ex: TerminalHardware]
  |                        |
  |                        +--rdf:type----> [ex:HardwarePlatform]
  |                        +--ex:displayWidth--> "320"
  |                        +--ex:displayHeight--> "200"
  |
  +--ccpp:component--> [ex:TerminalSoftware]
  |                        |
  |                        +--rdf:type----> [ex:SoftwarePlatform]
  |                        +--ex:name--> "EPOC"
  |                        +--ex:version-> "2.0"
  |                        +--ex:vendor--> "Symbian"
  |
  +--ccpp:component--> [ex:TerminalBrowser]
                           |
                           +--rdf:type----> [ex:BrowserUA]
                           +--ex:name--> "Mozilla"
                           +--ex:version-> "3.0"
                           +--ex:vendor--> "Symbian"
                           +--ex:htmlVersionsSupported--> [ ]
                                                            |
                           -------------------------------
                           |
                           +--rdf:type----> [rdf:Bag]
                           +--rdf:_1-----> "3.0"
                           +--rdf:_2-----> "4.0"
```

Figure 4.8 The CC/PP profile structure

CC/PP = Composite Capabilities/Personal Preferences; RDF = Resource Description Framework; HTML = Hyper Text Mark-up Language

contains one or more attributes from the vocabulary. Attributes in a schema are divided into components distinguished by attribute characteristics. In the CC/PP specification it is anticipated that different applications will use different vocabularies. According to the CC/PP framework a hypothetical profile might look like Figure 4.8.

A CC/PP schema is extended through the introduction of new attribute vocabularies, and a device capability profile can use attributes drawn from an arbitrary number of different vocabularies. Each vocabulary is associated with a unique XML namespace. This mechanism makes it possible to reuse attributes from other vocabularies. We should mention that the prefix *ccpp* identifies elements of the CC/PP namespace (component, property, structure, attribute) (**http://www.w3.org/2002/11/08-ccpp-schema**). The prefix *prf* identifies elements of the UAProf namespace (**http://www.wapforum.org/profiles/UAPROF/ccppschema-20020710**). The prefix *rdf* identifies elements of the RDF namespace (**http://www.w3.org/1999/02/22-rdf-syntax-ns**). The prefix *pss* identifies elements of the Packet Streaming Service (PSS) namespace (**http://www.3gpp.org/profiles/PSS/ccppschema-PSS5**). Attributes of a component can be included directly or may be specified by a reference to a CC/PP default profile. Resolving a profile that includes a reference to a default profile is time-consuming. When the PSS server receives the profile from a device profile server the final attribute values cannot be determined until the default profile has been requested and received. Support for defaults is required by the CC/PP specification.

Tools such as XMLSpy can be used to build RDF profile XML documents for different devices, although this will most likely be produced by the device manufacturer. An example RDF XML document for the Ericsson P800 is shown below, where various elements have been condensed in the interests of clarity. The XML document can be found at the following URL (**wap.sony ericsson.com/UAprof/P800R101.xml**):

```
- <rdf:Description ID="Profile">
  - <rdf:Description ID="HardwarePlatform">
     <prf:ScreenSize>208x320</prf:ScreenSize>
     <prf:Model>P800R101</prf:Model>
   - <prf:InputCharSet>
      ... [various character sets supported]
     </prf:InputCharSet>
     <prf:ScreenSizeChar>20x15</prf:ScreenSizeChar>
     <prf:BitsPerPixel>12</prf:BitsPerPixel>
     <prf:ColorCapable>Yes</prf:ColorCapable>
     <prf:TextInputCapable>Yes</prf:TextInputCapable>
     <prf:ImageCapable>Yes</prf:ImageCapable>
     <prf:Keyboard>OnScreenQwerty</prf:Keyboard>
     <prf:NumberOfSoftKeys>0</prf:NumberOfSoftKeys>
     <prf:Vendor>Sony Ericsson Mobile Communications</prf:Vendor>
   - <prf:OutputCharSet>
      ... [various character sets supported]
     </prf:OutputCharSet>
     <prf:SoundOutputCapable>Yes</prf:SoundOutputCapable>
     <prf:StandardFontProportional>Yes</prf:StandardFontProportional>
     <prf:PixelsAspectRatio>1x1</prf:PixelsAspectRatio>
     <prf:PointingResolution>Pixel</prf:PointingResolution>

- <rdf:Description ID="SoftwarePlatform">
     <prf:AcceptDownloadableSoftware>Yes</prf:AcceptDownloadableSoftware>
        ... [application/java support]
```

```
- <rdf:Description ID="NetworkCharacteristics">
    <prf:SecuritySupport>WTLS class 1/2/3/signText
- <prf:SupportedBearers>
    ... [ GPRS or CSD support]
- <rdf:Description ID="BrowserUA">
    <prf:BrowserName>Sony Ericsson</prf:BrowserName>
        ... [various wap browser support : frames, tables, wml, wmlscript, wmbp]
- <rdf:Description ID="WapCharacteristics">
    ... [ WAP support : PushMsgSize, WTAI, WMLScript, WMLDeckSize]

- <rdf:Description ID="MMSCharacteristics">
    <prf:MmsMaxMessageSize>300000</prf:MmsMaxMessageSize>
    <prf:MmsMaxImageResolution>640x480</prf:MmsMaxImageResolution>
  - <prf:MmsCcppAccept>
      <rdf:li>image/jpeg</rdf:li>
      <rdf:li>image/jpg</rdf:li>
      <rdf:li>image/gif</rdf:li>
      <rdf:li>image/png</rdf:li>
      <rdf:li>image/bmp</rdf:li>
      <rdf:li>image/x-bmp</rdf:li>
      <rdf:li>image/vnd.wap.wbmp</rdf:li>
      <rdf:li>application/smil</rdf:li>
      <rdf:li>application/x-sms</rdf:li>
      <rdf:li>audio/amr</rdf:li>
      <rdf:li>audio/wav</rdf:li>
      <rdf:li>audio/mid</rdf:li>
      <rdf:li>audio/x-wav</rdf:li>
      <rdf:li>audio/x-mid</rdf:li>
      <rdf:li>audio/midi</rdf:li>
      <rdf:li>audio/x-midi</rdf:li>
      <rdf:li>audio/basic</rdf:li>
      <rdf:li>audio/rmf</rdf:li>
      <rdf:li>audio/x-rmf</rdf:li>
      <rdf:li>text/plain</rdf:li>
      <rdf:li>text/x-iMelody</rdf:li>
      <rdf:li>text/x-eMelody</rdf:li>
      <rdf:li>text/x-vCard</rdf:li>
      <rdf:li>text/x-vCalendar</rdf:li>
      <rdf:li>text/x-vNote</rdf:li>
  - <prf:MmsCcppAccept-Charset>
      ... [various character sets supported]
    </prf:MmsCcppAccept-Charset>
    <prf:MmsVersion>1.0</prf:MmsVersion>
```

4.6.5   Terminal capability negotiation

WAP provides a mechanism to inform an origin server, such as the MMS relay/server, of the capabilities of the MMS user agent, which is known as the User Agent Profile (UAProf). It provides information about the characteristics of the display (e.g., size, colour support, bit depth), supported content types and network limitations (e.g., maximum message size). The UAProf data is encoded in an RDF data description language. It is conveyed, when the MMS user agent performs a WAP Session Protocol (WSP)/HTTP operation, such as a GET, to an origin server. It is up to the origin server to decode the RDF data, extracting any needed device characteristics, to guide the content generation or filtering operation it performs before returning data to the MMS user agent.

For the MMS, the MMSC should be able to utilize the capability information to make adjustments to the delivered MM contents. For example, an MMSC may delete a message component if the content type was not supported by the terminal. Alternatively, the MMSC may adapt an unsupported content type to adjust the size, colour depth or encoding format. WAP makes no requirements to the handling of these data or of any notifications that may be made to the user concerning such adjustments.

4.6.6 Digital Rights Management

There is currently limited support for Digital Rights Management (DRM) in the MMS handsets available in 2003. The MMS TS 23.140 release 5 (3GPP, 2002a) makes no specific mention of the topic, although proprietary support for simple forward-locking is available, to prevent a user from forwarding an MM to another user.

To protect copyrighted music or other content, Ericsson has made the T68i accept a special content header type. Here is how to protect a picture, for example:

```
Content-type: application/vnd.mms.ericsson.protected
X-Mms-Ericsson-Protected: image/gif; name="test.gif"
```

The protected entity cannot be forwarded from the phone, it can only play the sound or display picture. Chapter 7 on future recommendations discusses the initiatives proposed to tackle the issue of securing MMS content.

4.6.7 Provisioning

The quick and easy provisioning of MMSs is fundamental to their usability and has an obvious impact on the uptake. If a device is purchased without being configured to send and receive MMs then it is likely the user will never bother to enable this feature. If a change to a mobile operator's commercial MMS service requires a new device configuration, this could potentially affect millions of customers. The current solution is to preconfigure devices before they leave the shop or to configure devices Over The Air (OTA) afterwards, usually by allowing users to download configuration settings from the mobile operator's website.

While device provisioning is the most prominent configuration requirement, consideration should be given to configuring services and controlling the available services (Lannerstrom, 2002).

4.7 DEVICE PROVISIONING

As mobile operators and service providers currently need to be able to change settings for WAP and GPRS parameters remotely on the handset, they will need the possibility to change the MMS parameters and end-user preferences in that area on the SIM card and also in the handset device.

One of the major lessons learned from early WAP implementations is that services must be easy to use right from the start. Fast and easy configuration of the MMS settings is definitely one part of this. Provisioning of configuration settings, bound in the device or in the SIM, using an OTA method is a necessity.

MMSC configuration settings include a set of information elements needed to access network infrastructures. It is mainly a set of addresses for different access points related to MMS, such as:

- MMSC Uniform Resource Locator (URL) address (e.g., **http:// mms.o2.co.uk**);

- WAP gateway (for WAP implementations), IP address, port no., etc.;

- Access Point Node (APN) (e.g., **wap.o2.co.uk**).

Furthermore, there are other information elements that need to be configured – elements that depend on the provided service in terms of, for example, bearer, security, implementation protocol. Annex F in TS 23.140 (3GPP, 2002a) has a complete list of MMS configuration elements.

For many device manufacturers the OTA provisioning mechanism is proprietary, based around the transmission of an 8-bit binary SMS to a specific "port", related to an application on the phone that is able to prompt the user for acceptance of the configuration settings and add or overwrite these on the mobile phone (or SIM). The TS 23.048 specification (3GPP, 2002k) for configuring devices considers the wider issues of security and downloading applications for execution on the SIM card, in addition to its use for configuring services. It should be noted that this TS group does not specify MMS OTA.

For devices that support the Open Mobile Alliance (OMA) specification (WAP-189) for MMS settings configuration, the settings are written in XML and encoded in the WAP binary XML format for transmission to the device over WAP-PUSH/SMS with:

```
content-type: application/x-wap-prov.browser-settings
```

Once received they are then saved as a configuration message and can be used to update the settings on the device. It should be noted that we are still at early stages in the implementation of this functionality. In course of time, there will be a wide range of devices available in the market, some will have full MMS support and others will be more limited. Regardless of the capabilities of the subscriber's device at any given occasion, he or she may use a different device with a different potential the next time. However, configuring the device should not be an issue of concern for the subscriber; such a task should and often is a service offered by the mobile operator.

4.8 SERVICE PROVISIONING

Ease of use and familiarity regardless of device are two essential items for establishing and maintaining an MMS with a high degree of usage, and adding exciting new services must be managed to ensure a good customer experience. From a user perspective, the look and feel when activating the service should be the same regardless of what kind of device he or she is using at that moment. End-users need to be able to choose which services they want to use on their mobile device: a personalized set of services can be created and modified either via the Web or by using the mobile phone directly to search for and download new services. This level of personalization makes it easier to launch new services. Personalized services can help increase customer loyalty and reduce churn, at the same time reducing the need for customer care when the end-users perform service management themselves. An example of this level of personalization is the Trigenix platform from 3Glab and the Vodafone Live! service offering.

4.9 CONTROL PROVISIONING

Control provisioning includes mechanisms to be used for both device and service provisioning activities. It additionally covers control mechanisms during execution of services (i.e., while utilizing an application). Control mechanisms in this respect mostly relate to security aspects and, especially, authentication. Another important control aspect of MMS is its relation to DRM, focusing on who owns the rights to the content. Closely related to this comes the issue of charging for content and control mechanisms around this. Mechanisms for authentication and digital signatures can both be considered security services and typical elements of a commercial MMS offering:

- authentication (e.g., in order to reach an application or a service location);

- digital signature (e.g., when ordering a download of a billable content).

4.9.1 Device presentation

In general, the main difference between handsets is the level of support for different media formats. For example, some phones support the Adaptive Multi Rate (AMR) codec for recorded audio, some do not. Some phones support MIDI ring tones, some support Synthetic music Mobile Application Format (SMAF) and some support neither.

Other issues relate to the display size of images, as different phones have different display characteristics. There are also size limitations: different phones have different overall size limits for the size of MMS file that they support (ranging from 30 to 300KB in

handsets available in 2003). These differences are important given
that an MMS file is essentially a multipart message containing
different file content parts within the message.

   In the next subsection we will cover the issue of interoperability
with legacy handsets, how this is handled and the way content
adaptation can be used to overcome the differences between
MMS-capable handsets.

4.9.2   MMS and legacy handsets

Obviously, in order to use MMS the first thing all subscribers must
do is buy a new phone. The first phones to be fully capable of
creating and receiving MMS came onto the market in the second
quarter of 2002 – Ericsson's T68i colour phone and the Nokia 7650
Symbian-based phone with integral camera.

   In 2003 the penetration of MMS handsets in many European
markets has been low, because, even though the first MMS handsets have reduced in price, new models are normally relatively
expensive. It will take a couple of years before a substantial percentage of mobile subscribers are fully MMS-enabled. In order to encourage the early upgrade to MMS handsets, it has been necessary
for operators to provide some means for them to send messages to
people who do not have MMS handsets – at first, this will be
friends, family and colleagues. For this reason, many vendors
have implemented some form of "backwards compatibility" (i.e.,
MMSs can pass traffic to non-MMS handsets, downgrading the
content according to the limitations of the device). It is also possible
to send messages from MMS handsets to Internet email addresses,
and from the Web to MMS handsets. In this way a large addressable community, which is crucial to the uptake of any messaging
service, will be available to MMS users from the start. The MMS
specification does not indicate exactly how the sending of an MMS
message to "legacy" phones should work, and there is some variation in different vendor implementations. In its MMS platform, for
example, Comverse defines four levels of legacy support:

● SMS phone – the user receives an SMS notification and a URL,
   where the multimedia content can be viewed using a PC;

● WAP phone – the user receives an SMS notification and a URL, where the multimedia content can be viewed using the phone or a PC;

- WAP-PUSH phone (supporting WAP 1.2 or later) – the user receives a WAP-PUSH notification and a URL, with the option of retrieval using the phone or viewing with a PC;

- GPRS phone without MMS client – the user can receive the multi-media message and view this in a WAP browser, but with some content degradation (no animation or sound). There is no ability to send MMS messages.

The Nokia MMSC and Terminal Gateway will allow the users of MMS phones to send MMs not only to other MMS-capable phones but also to legacy GSM terminals. Users of legacy terminals receive an SMS notification and can connect to a website to view their MMS message.

4.9.3 Dynamic MMS content adaptation

MMSC support for dynamic content adaptation and conversion will be essential in offering content services that are interoperable across a range of devices. There are a number of differences between devices, some are outlined below:

- the Nokia 7650 can display a 176 × 144 image in GIF, JPEG, PNG or WBMP format, can support AMR audio and MIDI sound files and can handle MMS content up to 100KB in size;

- the Nokia 7210 can display the same image formats (but at 128 × 128), can support MIDI sound files and can handle content only up to 32KB in size;

- the Nokia 3510 can display a 96 × 65 image, as long as it is in monochrome GIF or WBMP format and does not provide any support for SMIL. The 3510 supports MIDI sound files and can handle content only up to 30KB in size.

And that is just looking at some of the differences in devices from a single manufacturer. The fact is that different devices are appearing all of the time from a variety of different manufacturers. This leads to the question of whether to develop content to the least common denominator and how to ascertain which device is being used to access content services?

Many MMSC products use the WAP/MMS UAProf capabilities to determine the MIME formats that a device supports as well as the maximum size of images supported by the device. Where required, the MMSC converts between common image formats (including but not limited to GIF, JPG, PNG, BMP and WBMP) to deliver an image supported by the device. For images larger than the maximum size supported by the device, the MMSC will automatically scale the image to fit the device, speeding up download times.

For audio formats, conversion between WAV and AMR is possible, but the WAV format is around 10 times the size of AMR. So, in many cases it is impractical to convert to WAV unless the device supports large MMS messages. MIME types not supported by the receiving device that cannot be converted should be removed prior to delivery to the receiving device to prevent compatibility issues and unnecessary download delays. Many content adaptation engines will dynamically update SMIL content before delivery to the mobile device, reflecting any necessary changes, or, if the device does not support SMIL, it removes the SMIL file, which helps by keeping the user from being confused by any references to "unsupported content types".

Although dynamic MMS content adaptation and conversion is a standard feature of many MMSC products, it is possible to use a MMS content adaptation proxy server, which would sit between the WAP gateway and an existing MMSC.

There is a limited potential security risk of new MMS viruses; this is due to the content being rendered within a SMIL player, which in itself is not an execution environment that would allow malicious code to be executed. There is still the potential of intentional or unintentional exploitation of bugs in various MMS client applications; we have already seen how large MMS messages can place the Nokia 3510 browser in an unstable state. Future updates to dynamic content adaptation and content conversion services may be put in place to block these exploits.

4.9.4 Synchronized Multimedia Integration Language

SMIL has been developed by the W3C as a way of animating and displaying video, text and audio content with a sequential element (e.g., a series of still images or a video clip made up from a series of images). One way of looking at the capability of SMIL is by comparison with the same basic principle used for conventional video editing standards. It divides a video clip into a sequence of time-coded frames, each of which has a start point and duration defined by the editor (Le Bodic, 2002). Until the late 1990s, it was only possible to produce this type of content with proprietary technologies, such as Macromedia's Shockwave and Real Networks' RealMedia. This changed in mid-1998 with the release of SMIL version 1.0, the first standards-based development to deliver a multimedia mark-up language. It should be noted that Microsoft, Real and Macromedia all supported the development of SMIL, with Real taking a leading role. The purpose of SMIL was to enable web developers to assemble multiple elements into a presentation slide deck (e.g., a video clip with audio, plus text and graphics). The current version of SMIL (version 2.0) was released in August 2001.

Two of the main drivers for the development of SMIL have been (1) the need to optimize the performance of streaming media over low-bandwidth Internet connections and (2) to remove the dependence on proprietary techniques. However, for MMS the importance of SMIL lies in its ability to control the sequential presentation of images and audio. It can be used, for instance, to make the frames of a comic strip appear one after the other, or to compile a birthday card comprising an audio clip playing over a cartoon sequence, or to construct a rolling deck of holiday snaps. SMIL is also important because it reduces the bandwidth needed for composite presentations in which a graphic is only one element (e.g., a graphic is transmitted only once and audio is transmitted as a separate object).

The OMA MMS conformance document specifies that MMSCs and handsets should include SMIL support. There are very few MMS handsets that do not support SMIL, the Nokia 3510 being one: it only has the ability to receive graphics, audio or text elements. The functionality available to developers of MMS content is shown in Figure 4.9.

The construction of an MM should be thought of as a slide deck. The slides we are talking about here are similar to a multimedia PowerPoint slide presentation. The MM slides are dynamic: you do not have to turn to the next slide by yourself, each slide has its own "show time", and after the defined period of "show time", the visible slide disappears and allows the next slide to appear and

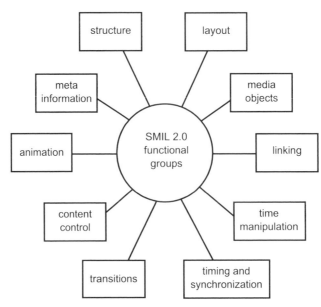

Figure 4.9 Functionality of SMIL

start its show. Not only are the slides dynamic but so are the objects (images, text, sounds): they can be set to appear at different time intervals and each object has its own "show time" as well. What makes the MM slides so dynamic and attractive? The power behind the presentation is the mark-up language SMIL, which is based on XML and is specified by the W3C. SMIL is used to define the layout of the MMS slides, timing of MMS slides and the multimedia objects on those slides. It is designed to lay out text, images and sounds, and to deliver streamed video. The SMIL used to compose an MM is just a subset of the whole language called MMS SMIL. The following SMIL code is from SonyEricsson's developer's guide, which we will use it to describe the MMS SMIL structure and functionalities:

```
<smil>
  <head>
    <meta name="title" content="vacation photos" />
    <meta name="author" content="Danny Wyatt" />
    <layout>
      <root-layout width="160" height="120"/>
      <region id="Image" width="100%" height="80" left="0" top="0" />
      <region id="Text" width="100%" height="40" left="0" top="80" />
    </layout>
  </head>
```

```
<body>
  <par dur="8s">
    <img src="FirstImage.jpg" region="Image" />
    <text src="FirstText.txt" region="Text" />
    <audio src="FirstSound.amr"/>
  </par>
  <par dur="7s">
    <img src="SecondImage.jpg" region="Image" />
    <text src="SecondText.txt" region="Text" />
    <audio src="SecondSound.amr" />
  </par>
</body>
</smil>
```

Useful resources:

http://www.helio.org/products/smil/tutorial/chapter1/index.html
http://www.realnetworks.com/resources/samples/smil.html
http://www.w3.org/TR/smil20/

4.9.5   SMIL structure overview

From the above example code you can see that SMIL is as simple as HTML. The tags <smil></smil> are just like <html></html> and the whole message is included inside these two tags. The head section defines both the brief information of the message and the layout of the message:

```
<meta name="title" content="Holiday photos" />
<meta name="author" content="Xin Guo" />
```

The <meta/> tag and its attributes provides brief information about the message. The "title" and "author" and their string contents tell the message receiver who made and sent the message and the message name. The functionalities of these two rows are similar to the "title" and "from" field of an email message.

Basically, every slide of a message should have the same layout, but different contents. The code between <layout> and </layout> tags is used to define the layout of all the slides of the message. Each slide is divided into two regions: image and text. For example, see the following code (see also Figure 4.10):

```
<root-layout width="160" height="120"/>
<region id="Image" width="100%" height="80" left="0" top="0" />
<region id="Text" width="100%" height="40" left="0" top="80" />
```

```
<layout>
        <root-layout width="128" height="128"/>
        <region id="Image" width="128" height="72" left="0" top="0" />
        <region id="Text" width="128" height="56" left="0" top="72" />
</layout>
```

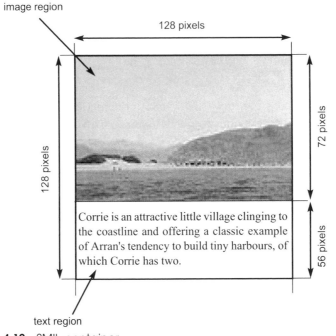

image region

128 pixels

128 pixels

72 pixels

Corrie is an attractive little village clinging to the coastline and offering a classic example of Arran's tendency to build tiny harbours, of which Corrie has two.

56 pixels

text region

Figure 4.10 SMIL container

The picture in this example was shot in the quaint village of Corrie, located on the wild coast of the island of Arran, Scotland

The <root-layout/> tag and its attributes define the area where the multimedia objects will be placed on the slide. Width="160" height="120″ means the slides will be displayed as 160×120 pixels on a PC screen, but different users have different mobile devices and display limitations. The message, once encoded, might not be properly displayed on some mobile devices: the current MMS syntax is designed to allow users to reformat the layout in the way best suited to their own devices. This is the reason that the current implementations of MMS terminals only allow one image region and one text region per slide. A customer device may choose to replace any incoming layout information with its own fixed layout (e.g., one that it uses for all MMS messages regardless of specified layouts):

```
<region id="Image" width="100%" height="80" left="0" top="0" />
```

In the above code the <region/> tag and its attributes define the area of an image region or a text region and the id attribute is the name of the region. Make sure the areas you assign to both regions are no larger than the area you defined in the <root-layout/> tag:

```
<body>
  <par dur="8s">
    <img src="FirstImage.jpg" region="Image" />
    <text src="FirstText.txt" region="Text" />
    <audio src="FirstSound.amr"/>
  </par>
  <par dur="7s">
    <img src="SecondImage.jpg" region="Image" />
    <text src="SecondText.txt" region="Text" />
    <audio src="SecondSound.amr" />
  </par>
</body>
```

The above code defines the body section of the actual slides. The <par> and </par> tags define the appearance and timing of the multimedia objects inside them. ''Par'' means parallel and every multimedia object inside the <par> and </par> tags is shown simultaneously <par> and </par> are counterparts to <seq> and </seq> tags, which make the multimedia objects display in a sequence). The ''dur'' attribute defines how long the slide will show. The <img/>, <text/> and <audio/> (Figure 4.11) tags define which object is being referred to and the region and sources of the objects. Note that the image region must contain the image object and the text region must contain the text object (this can be different in general SMIL).

Table 4.1 shows the MMS SMIL tags and their functionalities.

4.9.6 Application development

There are a number of application SDK (Software Development Kit) environments for the development and viewing of MMS content. Some of these extend to include simulated delivery to phones through an MMSC using a proprietary implementation of MM7.

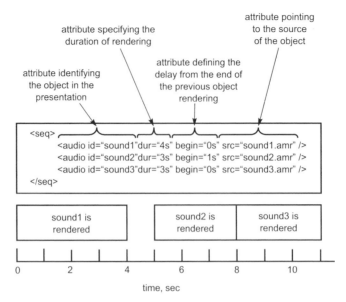

Figure 4.11 SMIL sound sequence

In the next few sections we will discuss the Nokia and Ericsson application development environments to outline the key features and capabilities available within them. We have included some practical suggestions to overcome some of the difficulties in developing MMS content.

4.9.7 Nokia MMS development environment overview

Nokia has provided the following tools to create a new development environment:

- Nokia Developer's Suite for MMS;
- Adobe GoLive 6.0;
- Nokia MMSC External Application Inter Face (EAIF) emulator;
- Series 60 Content Authoring SDK 1.0 for Symbian OS (Nokia edition).

Nokia Developer's Suite for MMS integrates with Adobe GoLive 6.0 and adds new features to enable MMS message creation (Nokia, 2002a). This provides the developer with MMS encapsulation (the process of converting a SMIL presentation into an MMS message), emulator integration with the MMS terminal emulator and the EAIF emulator. It also supports deployment of the created MMS message to an emulator (including MMS phone emulators and Nokia Series 60 MMS SDK for Symbian OS).

4.10 ADOBE GOLIVE CONTENT AUTHORING

MMS content development using Adobe GoLive begins by dragging an object icon from the palette to the editing window. After you drop it, the related element tags will appear in the "source panel" (Figure 4.12). Then enter the attributes, names and values (such as filename of image, sound file and text file). After entering this information and performing an automated check of the syntax, correct any errors and then check the syntax again until it reports "No errors detected"; this can then be saved as an SMIL presentation.

To create an MMS message use the Adobe GoLive Nokia Tools menu to convert the SMIL presentation to an MMS message. As an example, open an SMIL presentation. Then, after creating your MMS message, select MMS Encapsulation and the MMS header dialog window will appear and the information for the MMS headers will need inputting. Enter the sender's email address or phone number in the "Form" field, enter the receiver's email address or phone number in the TO, CC or BCC fields (one of these must be populated). When entering a phone number in these fields, note that this requires "/TYPE=PLMN" (Public Land Mobile Network) after the end of the phone number. After entering all the information, click "Encapsulation", when a window pops up to indicate that the MMS encapsulation has been completed successfully and the SMIL file has been converted to an MMS message. As an MMS message is normally larger than the SMIL presentation, select Message information if you want to check the MMS message size.

Now that a new MMS message has been created, Adobe GoLive's Nokia Tools can be used to send the MMS message to the EAIF emulator: Series 60 MMS SDK for Symbian OS, or another phone simulator that can receive MMS messages. We will describe how to send MMS messages in the following sections.

Table 4.1 SMIL tags

Tag	Attributes		Functionalities
<smil> </smil>	None		Specifies the SMIL language
<head> </head>	None		Contains the metadata and layout of the message
<body> </body>	None		Contains the body of the message and encloses the actual slides
<meta/>	Name	the name of the metainformation	Tag for metainformation header to be put in the message
	Content	the actual content of the metainformation	
<layout> </layout>	None		Specifies the layout of all the slides in a message
<root-layout/>	Width	width of the entire area	Specifies the whole area the message will fill. The maximum area is 160×120 pixels
	Height	height of the entire area	
<region/>	Id	the name of the region	Defines a region of a slide
	Top	the top of the region (in pixels)	
	Left	the left edge of the region	
	Width	width in pixels	
	Height	height in pixels	
	Fill	scale width or height to fill space	
	Hidden	hide the right and bottom sides if the object is larger than the region or fill in the space if the object is smaller than the region	
	Meet	scales the object up, preserving its aspect ratio, until either its height or width fits the height or width of the region	

Tag	Attributes		Functionalities
	Slice	scales the object down, preserving its aspect ratio, until either its height or width fits the height or width of the region	
<par> </par>	Begin	the beginning time of the slide	Defines parallel execution of objects contained within these tags
	End	the end time of the slide	
	Dur	the duration of the slide	
<seq> </seq>	Begin	the beginning time of the slide	Defines sequential execution of objects contained within these tags
	End	the end time of the slide	
	Dur	the duration of the slide	
</img/>	Src	source of the image	Defines the images displayed on the slide
	Region	which region the image is placed in (the image region in MMS SMIL)	
	Alt	alternative text for image	
<text/>	Src	the source of the text	Defines the text displayed on the slide
	Region	which region the text is placed in (in MMS SMIL the text object is in the text region)	
	Alt	alternative display text	
<audio/>	Src	the source of the audio file	Defines the sounds to play when the slide is on show
	Alt	the alternative text for audio	

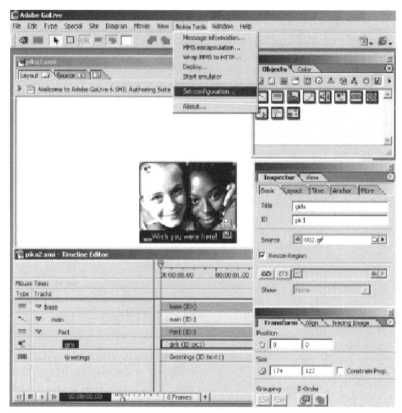

Figure 4.12 Adobe GoLive SMIL development environment
Source: Adobe/Nokia

4.11 SERIES 60 CONTENT AUTHORING SDK FOR SYMBIAN OS

Series 60 Content Authoring SDK 1.0 for Symbian OS can be installed independently of any supporting application. You can use the Series 60 MMS emulator to preview an MMS message without any supporting application. However, there are two supporting applications that can work with the Series 60 Content Authoring SDK:

• Nokia Mobile Internet Toolkit (NMIT) version 3.1;

• Adobe GoLive 6.0 (integrated with Nokia Developer's Suite for MMS).

NMIT is used for developing mobile Internet content using XHTML /WML (Wireless Mark-up Language) and MMS messages. Like Adobe GoLive, it integrates with Nokia Developer's Suite for MMS. NMIT can be used to create MMS messages and deploy the MMS messages to the Series 60 MMS emulator.

4.11.1 Using the Series 60 Content Authoring SDK for Symbian OS

For PC-based testing of MMS content, the Series 60 platform provides an MMS emulator that allows developers to preview the MMS message they have created. It is possible to send MMS messages to the Series 60 MMS emulator. The Series 60 Content Authoring SDK has many other features, but the focus of this section is on usage of the Series 60 MMS emulator. It can receive, edit and send MMS messages and developers can use it to preview the MMS messages they have created before sending them to a physical Series 60 MMS phone. As mentioned, the Series 60 MMS emulator can be used with or without a supporting development environment. To use an external content development tool with the Series 60 MMS emulator, either Adobe GoLive or NMIT will require further configuration.

4.11.2 Sending and receiving an MM with the Series 60 MMS emulator

Using the Adobe GoLive and Nokia MMS development suite, MMS messages can be sent and then received by the Series 60 MMS emulator. The sending of messages is essentially a deployment of the encapsulated MMS file to a local working directory where the MMS emulator is able to retrieve and display the MMS content. Ensure the ''Path to the emulator'' variable is correctly set:

```
C:\Nokia\Devices\Series_60\Epoc32\Release\wins\UREL\EPOC.EXE
```

Also ensure the ''Path to the deployment folder'' variable is set to the ''mmsin'' folder:

```
C:\Nokia\Devices\Series_60\Epoc32\Wins\c\mmsin
```

Figure 4.13 Adobe GoLive SMIL example page using Symbian emulator
Source: Adobe/Nokia

The "deploy message" box pops up to indicate that the MMS message has been successfully deployed to the Series 60 MMS emulator's "mmsin" folder. Now that the MMS message has been sent to the Series 60 emulator, it can be viewed from Adobe GoLive by selecting "Start emulator" from the Nokia "Tools menu" in Adobe GoLive. The emulator then loads the MMS message and displays it.

4.12 NOKIA MMS JAVA LIBRARIES

Nokia has provided MMS Java libraries to help MMS developers to create external applications to the MMSC. The libraries contain sample code, presenting examples of how to create, send, receive and decode MMs. Developers can learn the basic functionalities that are needed to develop external applications; the sample code can also be used in a commercial run-time. The examples provided by the MMS libraries give the following basic functionalities:

- Creating and encoding MMS messages – Nokia MMS Java libraries provide an example of how to create and encode an MMS message from different contents. They show how message creation can be done according to the WAP-209 MMS encapsulation specification. Developers can use this message creation example when they develop originating or filtering applications.

- Decoding MMS messages – Nokia MMS Java libraries provide an example of how to decode received messages according to the WAP-209 MMS encapsulation specification and how to extract multimedia contents from the message body as well. Developers can use this message decoding example when they develop terminating or filtering applications.

- Sending MMS messages to the MMSC – HTTP 1.1 protocol EAIF is used when sending MMS messages between an application and an MMSC. Nokia MMS Java libraries provide an example of how to send an MMS message to the MMSC according to this proprietary specification (see the section on MM7 VASP [Value Added Services Provider] on p. 134 for more information).

4.12.1 Installing Nokia MMS Java libraries

Before installing the Nokia MMS Java libraries, install JRE 1.3.1 or a higher version (JRE can be downloaded from **http://java.sun.com**). After installing the Nokia MMS Java libraries a folder named "mmslibrarysdir" will have been created, which contains the following folders: /doc, /mmslibrary, /samples, /src. The folder *mmslibrary* contains the .jar file. Further modify the CLASSPATH by adding the following path:

```
mmslibrarydir/mmslibrary/MMSLibrary.jar
```

4.12.2 Ericsson MMS development environment overview

Ericsson provides the following tools for MMS development:

- SonyEricsson MMS Composer version 1.1;

- SonyEricsson MMS Home Studio AMR converter.

The SonyEricsson MMS Home Studio (Figure 4.14, top) has similar functionalities to the MMS Composer (Figure 4.14, bottom), but our main focus here is on the SonyEricsson MMS Composer (Ericsson, 2002).

4.12.2.1   SonyEricsson MMS Composer version 1.1

The SonyEricsson MMS Composer version 1.1 can be downloaded from Ericsson's website at (**www.ericsson.com/mms**). A developer is initially presented with a workspace containing four buttons and a viewing area. The buttons control the addition or reviewing of content (namely, images, sounds, play and browse). A SMIL presentation is composed of several slides, and use of the MMS composer allows the creation of slides and timing of the slides, which can be used to give the appearance of animation. A slide is composed of a "slide container" and several objects placed in the container. Each object can be an image, text or sound clip. Adding image and sound objects to each slide container is accomplished by dragging the object from the viewing area onto the container. A container can only hold one object of one type. Adding a second image object onto the container will replace the previous image object. A "time section" is then created and placed in the time line, which consists of a time period for the display of an image along a pink track. The time period for playback of any sound associated with the slide container is given a purple track. Using the mouse to lengthen or shorten either side of each coloured time track will change the starting point and duration of an object, the time an image appears or the duration of audio playback.

Finally, the SDK allows for MMs to be transferred to and from an Ericsson phone using a cable or Bluetooth.

4.13   AUDIO CONVERSION

Where audio files for MMS messages are required in AMR codec format and the existing content is stored as a WAV file, then

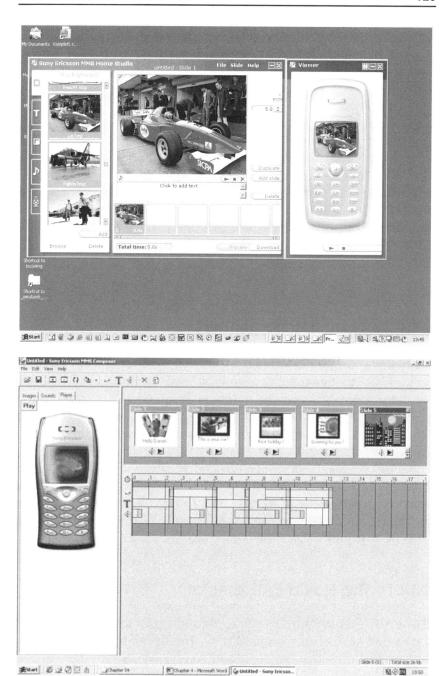

Figure 4.14 Ericsson SDK development environment. Top: Home Studio; bottom: MMS Composer

Source: Ericsson

conversion is required. An AMR file is small in comparison to the WAV equivalent. There is a AMR converter that can be down-loaded from the Ericssion website. In order to convert a file, the following executable command line can be invoked with the following syntax:

```
converter wav2amr filename.wav filename.amr MRxxx
```

The "filename" is the actual filename string and the "xxx" repre-sent a three-digit number: 4.75, 5.15, 6.70, 7.4, 7.95, 10.2, 12.2. These represent the coding rate (4.75 to 12.2kbps) that the AMR codec supports. After this process the generated .amr files are stored in the converter folder. Conversion can be made from .amr files to .wav files in a similar way.

4.14 APPLICATION TESTING

There are a number of simulators, message compilers and notifica-tion utilities that enable the testing of MMS applications. The EAIF emulator from Nokia is an interface designed to test application delivery of MMs. Argogroup supply a test suite called Monitor Master, a quality management platform that tests, monitors and reports on the quality and usability of content on WML, SMS, Enhanced Message Service (EMS) and MMS-enabled devices.

In this section we will explore the Nokia test emulator for three delivery scenarios and discuss the technical implementation of compiling SMIL content and notification. The section concludes with an overview of the MM7 interface.

4.14.1 The Nokia EAIF emulator

The EAIF enables third-party application providers' MMSs to be developed against the Nokia MMSC. The EA (external application) and service provider can send and receive MMS messages to and from the MMSC via EAIF (Nokia, 2002c). The EAIF is not a stan-dardized interface, and some functions of EAIF are the property of Nokia. 3GPP is responsible for MMS standardization (discussed in the section on MM7 VASP on p. 134). EAIF uses HTTP 1.1 and persistent connections. It also uses the POST method to carry MMS message and delivery reports to and from the MMSC.

4.14.1.1 External application

There are two ways for MMS developers to create MMS applications: they can either develop the application from the client-side (on the mobile terminal) or externally from the MMSC. Client-side applications depend on the capabilities of the device (this has already been discussed in the previous section on "application testing"). An EA is an application that interfaces to the MMSC. An MMSC treats messages received from such applications in a similar way to messages sent by an end-user. There are three types of EA: originating, terminating and filtering.

4.14.1.1.1 Originating case

In the originating application case, the application is acting as a Web client originating the messages. Correspondingly, the MMSC is acting as a Web server receiving the HTTP requests from the client (application). Figure 4.15 illustrates the roles of the application and Java classes. When sending an MMS message to the MMSC, Java classes such as those in Figure 4.15 present ways in which the following functions can be implemented:

- Message encoding to the format specified in WAP-209 for MMS encapsulation and message content is encapsulated according to MIME.

- message delivery to the MMSC is over EAIF and the messages are delivered over HTTP.

4.14.1.1.2 Message response reception from the MMSC

The Java classes present ways in which to encapsulate the MMS content (images, text, etc.) using MIME and to create and encode the content that is contained in a MMS Protocol Data Unit (PDU). After message creation, the Java classes present ways in which to carry the encoded message, using the HTTP POST request to deliver the message over the EAIF to the MMSC. The Java classes in Figure 4.15 also provide an example of response reception from the MMSC.

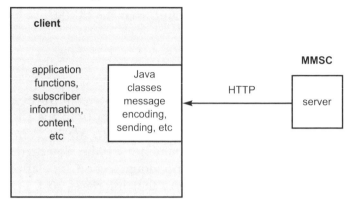

Figure 4.15 Originating application use case
HTTP = Hyper Text Transfer Protocol; MMSC = Multimedia Messaging Service Centre

4.14.1.1.3 Terminating case

In the terminating application case, the MMSC is acting as a Web client originating the messages. Correspondingly, the application is acting as a Web server receiving the MMs in the HTTP requests. Figure 4.16 illustrates the roles of the application and Java classes in the terminating application case. In the message reception from the MMSC, the Java classes illustrate how the binary-encoded and MIME-encapsulated MMS messages can be decoded. Such other terminating application functions as message reception from the MMSC are not presented in the Java classes.

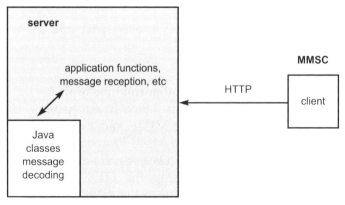

Figure 4.16 Terminating application use case
HTTP = Hyper Text Transfer Protocol; MMSC = Multimedia Messaging Services Centre

4.14.1.1.4   Filtering case

In the Nokia EAIF emulator the filtering application functionality is a combination of terminating and originating application function- alities: the message can be sent to the application as in the terminat- ing application case and received from the application as in the originating application case.

4.14.2   Creating and compiling MMS message files

As the MMS message file format is a binary file format, special tools are required to create MMS message files: the MMSCOMP utility is provided to assist in the creation of MMS message files and accepts text input files to create a binary MMS Message file. No standard format has been defined for a text version of an MMS message file; however, a format can easily be derived based on the MMS encapsulation protocol specification (WAP-209). The MMSCOMP utility accepts as input a file that contains text representations of the MMS header and one or more files (image, sound, text, etc.) to comprise the multipart message content. MMSCOMP is a command line utility from **www.nowmms.com** that accepts the following command line format:

```
MMSCOMP [-ccharset] header.file [data1.file [data2.file [data3.file ...]]]
```

-ccharset is used to specify a character set for the text components of the input files; hence this parameter is optional. As a matter of interest, the default character set is iso-8859-1. Other supported character sets include big5, iso-10646-ucs-2, iso-8859-1, iso-8859-2, iso-8859-3, iso-8859-4, iso-8859-5, iso-8859-6, iso-8859-7, iso-8859-8, iso-8859-9, shift_JIS, us-ascii and utf-8. A header.file is a text file that contains text representations of the MMS message header. Supported MMS message headers include:

```
X-Mms-Message-Type: m-retrieve-conf
                                                    (required)
X-Mms-Transaction-Id: text-string
X-Mms-Version: 1.0
Message-Id: text-string
                                            (usually x@x format)
Date: HTTP-date-format
From: address@domain or +InternationalPhoneNumber/TYPE=PLMN
                                (Address-present-token is assumed)
```

```
To: address@domain or +InternationalPhoneNumber/TYPE=PLMN
                          (use multiple headers for multiple recipients)
Cc: (same format as To)
Bcc: (same format as To)
Subject: text-string
X-Mms-Message-Class: Personal, Advertisement, Informational or Auto
                                              (default is Personal)
X-Mms-Priority: Low, Normal or High
                                               (default is Normal)
X-Mms-Delivery-Report: Yes or No
                                                  (default is No)
X-Mms-Read-Reply: Yes or No
                                                  (default is No)
Content-type: MIME-Type
                  (default is application/vnd.wap.multipart.related,
                               override default with caution!)
X-NowMMS-Content-Location: filename;content-type
                  (optional, use multiple headers for multiple files)
```

Only the X-Mms-Message-Type header is required, other headers are optional. It is recommended that "From" and "Subject" headers be always included. Note that while the message may contain multiple recipients in the "To", "Cc" and "Bcc" headers, the gateway itself will only send the MMS notification message to one recipient at a time, as specified in the "PhoneNumber" parameter passed in a URL request.

At least one data file must be specified to provide the content of the MMS message. This data file can be specified on the command line (e.g., data1.file, data2.file, data3.file, . . .), or it may be specified in the MMS header file with one or more X-NowMMS-Content-Location headers. If the first data file is a SMIL file, then MMSCOMP will automatically parse all "src" references in the SMIL file and include any referenced files in the MMS multipart message file automatically. If a SMIL file is to be included for presentation of the MMS message, it is recommended that it always be specified as the first data file to the MMSCOMP command.

MMSCOMP determines the MIME type of each file based on the file extension. Alternatively, when using the "X-NowMMS-Content-Type" header, the content type can be specified following the file name. File extensions of .jpg, .jpeg (image/jpeg), .gif (image/gif), .txt (text/plain), .wbmp (image/vnd.wap.wbmp) and .smil (application/smil) are recognized automatically. Other file extensions are read from the MMSCTYPE.INI file, or the Windows registry, under the registry key HKEY_CLASSES_ROOT\.extension, where ".extension" is the extension of the file.

The output of the MMSCOMP command will be stored in a file that matches the name of the input header file, but with ".MMS" as the file extension. For example, assume that:

1. You have created an MMS message header file named "test.hdr".

2. You have created a SMIL file named "testfile.smil", which references three external files through the following references in the SMIL file:

```
<img src="image.jpg" region="Image"/>
<audio src="sound.amr"/>
<text src="text.txt" region="region1_1"/>
```

3. The "image.jpg", "sound.amr" and "text.txt" files referenced by the "testfile.smil" file are located in the same directory as the "testfile.smil" file.

To create a binary MMS file, run:

```
MMSCOMP test.hdr testfile.smil image.jpg sound.amr text.txt
```

If you want to specify a character set for the text file, include the ccharset parameter:

```
MMSCOMP -cUTF-8 test.hdr testfile.smil
```

The output of the MMSCOMP file will be "test.mms" (e.g., the same filename as "test.hdr", but with a ".mms" file extension). To send the compiled MMS file, you can store the file on a Web server with a file extension .mms and ensure it is configured to use a MIME content type of "application/vnd.wap.mms-message", then use a WAP Push Proxy Gateway (PPG) to send an MMS notification message to an MMS User Agent (UA) on a mobile device.

The MMSCOMP utility works in conjunction with tools such as SonyEricsson's MMS Composer, which creates the SMIL files that are often used in MMS messages, but it does not create the complete binary MMS message file. To use output from SonyEricsson's MMS Composer, use "File/Export" to export your MMS message to a specified directory. The Composer will output an SMIL file and any included MMS message components to the directory specified. Then create an MMS message header file and run MMSCOMP, passing the name of the MMS message header file and the name of the SMIL file output by SonyEricsson's MMS Composer.

4.14.3   Sending MMS notifications

As we have seen, MMS messages are sent using a combination of SMS and WAP technologies. When an MMS message is sent, a mobile device receives an MMS notification message via SMS. When this MMS notification message is received by the mobile device, the mobile device automatically initiates a WAP gateway connection to download the content of the MMS message. To send an MMS message, you must first create an MMS message file. The format of an MMS message file is documented in the MMS encapsulation protocol specification published by the OMA (WAP-209) (described in the previous section). The MMS message file format consists of an MMS message binary header, followed by a multipart MIME message where the multipart message is encoded in a binary multipart format as defined by WSP specification. This binary MMS message file is stored on a Web server using a MIME type of application/vnd.wap.mms-message and an MMS message type of m-retrieve-conf. A subset of the binary MMS header is sent as an MMS notification message (MMS message type m-notification-ind) via SMS to the mobile device together with a URL pointer to the location of the complete message.

Once an MMS message file has been built and published via a Web server, the MMS notification message can be sent using SMS from a PPG. Using the test_ppg application supplied with the open source Kannel gateway, communication with a PPG can be made to send an SMS containing the WAP-PUSH notification in order to deliver an MMS message to the user:

```
../test/test_ppg -e base64 -a mms -c mms
"http://ppg.host.com:8080/wappush?username=yyy&password=xxx"
m-notification-ind.txt pap.txt

FILE: m-notification-ind.txt

X-Mms-Message-Type: m-notification-ind
X-Mms-Transaction-Id: 125
X-Mms-Version: 1.0
X-Mms-Message-Class: Personal
X-Mms-Message-Size: 4910
X-Mms-Expiry: 256; type=relative
```

```
X-Mms-Content-Location: http://192.168.1.1/mmstest/test.mms
X-WAP-Application-Id: x-wap-application:mms.ua

FILE: pap.txt
<?xml version="1.0"?>
<!DOCTYPE pap PUBLIC "_//WAPFORUM//DTD PAP//EN"
"http://www.wapforum.org/DTD/pap_1.0.dtd">
<pap>
<push-message push-id="9fjeo39jf084 N host.com"
progress-notes-requested="false">
<address address-value="WAPPUSH=+447740305115/TYPE=PLMN N ppg.
                                              carrier.com">

</address>
</push-message>
</pap>
```

4.15   APPLICATION DEPLOYMENT

In this section we will discuss the requirements for deploying MMS applications, which broadly covers the following areas:

- content management;

- the VASP MM7 interface.

4.15.1   Content management

There are a number of developer and composer tools available (e.g., SonyEricsson's Composer and Adobe GoLive). These tools are very useful for learning about the constructs for building SMIL presentations; however, each device may display content in different ways, or be unable to display content at all if the device does not support the graphic format/size or audio format. Content management systems must enable publishers to be connected with operators through the MM7 interface to enable billing. The output of MMS content should be optimized for each device, and that content should be stored only once in the system.

The middle panel of Figure 4.17 shows how image transcoding and delivery works. There are at least 50 different types of MMS phones, so it would unmanageable to create 50 different versions of MMS content. Therefore, a system with automatic transcoding is needed. The content to build an MMS is stored in a database in XML format and will be transcoded first when sent to the end

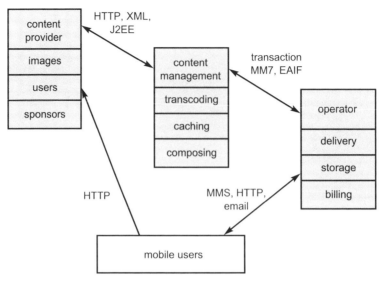

Figure 4.17   Content management architecture

HTTP = Hyper Text Transfer Protocol; XML = Extensible Mark-up Language; J2EE = Java 2 Enterprise Edition; MM7 = interface to VASP; EAIF = External Application Inter Face; MMS = Multimedia Messaging Service

device (optimized in size, colour, audio, SMIL and other features). A media archive is used to support the storage and preview of video content, which is stored in multiple formats, including H.263 or MPEG4. Such a media archive presents the opportunity to preview content.

4.15.2   Value Added Service Provider

MMS VAS applications provide Value Added Services (VASs) to MMS users. In many ways MMS VAS applications behave like a fixed MMS user agent. However, MMS VAS applications may provide some additional features like MM recall between MMS VAS applications and the MMS relay/server that are not available to MMS user agents (Hietala, 2002). The 3GPP specification does not cover what kind of applications might be available and how the MMS VAS application should provide these services.

MMS VAS applications may need to generate Charge Detail Records (CDRs) when receiving MMs from the MMS relay/server and when submitting MMs to an MMS relay/server. The interaction between an MMS relay/server and the MMS VAS application should be provided through the MM7 interface, to support the following:

- authentication;

- authorization;

- confidentiality;

- charging information.

As mentioned on p. 126, Nokia introduced an implementation of the MM7 interface for the Nokia MMS solution called EAIF. MM7 is a mobile Web service interface standardized by the 3GPP. In the Internet, Web service interfaces have rapidly become a widely accepted business integration technology, and in the mobile environment will act as a bridge between mobile network servers and web/application servers, a way of integrating the Internet and mobile domains (Harris, 2002).

Common standard-based Web service interfaces will provide operators and content developers with opportunities to share new components and support quick and cost-effective deployment of revenue-generating mobile services. For instance, an MMS content provider and an operator can integrate their systems using a Web service interface, so that the MMS content, such as a daily sports update, will be made available to mobile users. Mobile Web service interfaces will allow operators to increase their revenues by offering their unique mobile assets – such as billing capabilities, location or presence information – in addition to transport, to any service provider that would benefit from such services.

Mobile Web service interfaces are standardized in various industry fora, such as the OMA, Third Generation Partnership Project (3GPP) and the W3C, and will be implemented based on widely accepted standard technologies, thus providing a common and interoperable way to define, publish and use mobile VASs. These technologies are SOAP, XML, WSDL (Web Services Description Language) and HTTP.

SOAP and XML have been selected by 3GPP as the standard for the Open Services Architecture (OSA). They are currently being standardized by W3C, and use an XML data format for platform-independent data representation. Combined with XML templates,

SOAP and XML enable a single source of content development to be deployed on a range of terminals, including PC, mobile and PDA. SOAP and XML are transport-neutral and HTTP is the standard transport, thus enabling the widest interoperability with low barriers to entry.

A VASP can use SOAP messages as HTTP POST method bodies. The destination URLs are well known (preconfigured) or directory-enabled and SOAP processing is simple, requiring only XML and HTTP libraries. It also supports the standard HTTP authentication mechanisms. The MMS Application Programming Interface (API) for third-party providers must comply with the OSA general framework TS 23.127 (3GPP, 2002c). The OSA framework enables such network service capabilities as authentication and discovery. The OSA API insulates applications from changes in the network capabilities: when new services are enabled, existing applications continue to work. A lot of the attention of MMS developers is focused on the MM7 standard, because it defines how they will use SOAP messages to send MMS messages to consumers via an MMSC, and several MMSC vendors have committed themselves to delivering MM7. There are other proprietary mechanisms that could be used to deliver VASP messages:

1. HTTP using a POST with the "multipart/form-data" MIME content type. One example is the use of a Web form that uploads an MMS file containing this content type from a Web browser.

2. SMTP – where you send a message to **phonenumber@mms.domain** and include the MMS message content as email message attachments.

3. If the MMS content is precompiled and resides on an external Web server, a simple HTTP GET can be issued to ask a server to send out an MMS notification.

4.15.3 Technical specification of how VASP applications are delivered with MM7

The MMSE may support VASs in addition to the basic messaging services defined for MMS. These VASs may be provided by the

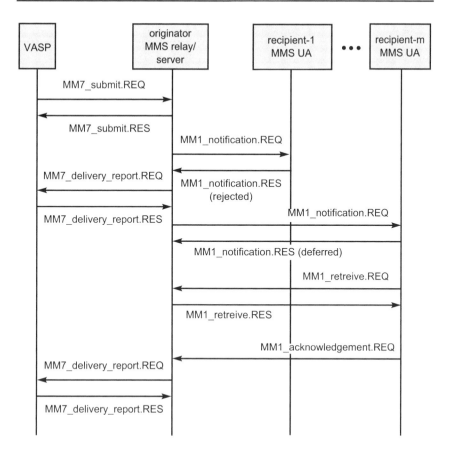

Figure 4.18 VASP message submit

VASP = Value Added Service Provider; MMS = Multimedia Messaging Service; UA = User Agent

network operator of the MMSE or by third-party VASPs. This section defines the interworking between the MMS relay/server and the VASP.

4.15.3.1 Submitting a VAS MM

Figure 4.18 outlines the operations necessary for a VASP to provide the service by sending an MM to one or more subscribers or to a distribution list. The abstract messages involved are outlined in Tables 4.2–4.5 from type and direction points of view.

Table 4.2 Information elements in the MM7_submit.REQ (Request)

Information element	Presence	Description
Transaction ID	Mandatory	Identification of the MM7_submit.REQ/ MM7_submit.RES pair
Message type	Mandatory	Identifies this message as an MM7_submit request
MM7 version	Mandatory	Identifies the version of the interface supported by the VASP
VASP ID	Optional	Identifier of the VASP for this MMS relay/server
VAS ID	Optional	Identifier of the originating application
Sender address	Optional	The address of the MM originator
Recipient address	Mandatory	The address of the recipient MM. Multiple addresses are possible or the use of an alias that indicates the use of a distribution list. It is possible to mark an address to be used only for informational purposes
Service code	Optional	Information supplied by the VASP that may be included in charging information. The syntax and semantics of the content of this information are beyond the scope of this specification
Linked ID	Optional	This identifies a correspondence to a previous valid message delivered to the VASP
Message class	Optional	Class of the MM (e.g., advertisement, information service, accounting)
Date and time	Optional	The time and date of the submission of the MM (time stamp)
Time of expiry	Optional	The desired time of expiry for the MM (time stamp)
Earliest delivery time	Optional	The ealiest desired time of delivery of the MM to the recipient (time stamp)
Delivery report	Optional	A request for delivery report (Boolean – true or false)
Read reply	Optional	A request for confirmation via a read report to be delivered (Boolean – true or false)

Information element	Presence	Description
Reply-charging	Optional	A request for reply-charging
Reply deadline	Optional	In case of reply-charging the latest time of submission of replies granted to the recipient(s) (time stamp)
Reply-charging size	Optional	In case of reply-charging the maximum size for reply MM(s) granted to the recipient(s)
Priority	Optional	The priority (importance) of the message
Subject	Optional	The title of the whole MM
Adaptations	Optional	Indicates whether VASP allows adaptation of the content (default is "true")
Charged party	Optional	An indication of which party is expected to be charged for an MM submitted by the VASP (e.g., the sender, receiver, both parties or neither)
Content type	Mandatory	The content type of the MM
Content	Optional	The content of the MM
Message distribution indicator	Optional	If set to "false" the VASP has indicated that the content of the MM is not intended for redistribution. If set to "true" the VASP has indicated that the content of the MM can be redistributed

4.15.3.2 Delivery request

This message sequence diagram addresses cases where messages are passed by the MMS relay/server to a VASP for processing (e.g., cases where the message originated from the MMS UA (Figure 4.19).

4.15.3.3 Cancel and replace of MM

A VASP can control or change the distribution of a message. This operation will allow the VASP to cancel a submitted message

Table 4.3 Information elements in the MM7_submit.RES (Response)

Information element	Presence	Description
Transaction ID	Mandatory	The identification of the MM7_submit. REQ/MM7_submit.RES pair
Message type	Mandatory	Identifies this message as an MM7_submit response
MM7 version	Mandatory	Identifies the version of the interface supported by the MMS relay/server
Message ID	Conditional	If status indicates success, then this contains the MMS relay/server-generated identification of the submitted message. This ID may be used in subsequent requests and reports relating to this message
Request status	Mandatory	Status of the completion of the submission, no indication of delivery status is implied
Request status text	Optional	Text description of the status for display purposes, qualifying the request status

prior to delivery or replace a submitted message with a new message.

4.15.3.4 Delivery reporting to VASP

This part of MMS covers the generation of a delivery report from the MMS relay/server to the VASP.

4.15.3.5 Generic error handling

When the MMS relay/server or VASP receives an MM7 abstract message that cannot be replied to with a specific response, it replies using a generic error message as described here. To get a correlation between the original send request and the error response, every abstract message on the MM7 reference point includes a transaction ID.

Table 4.4 Information elements in the MM7_deliver.REQ

Information element	Presence	Description
Transaction ID	Mandatory	Identification of the MM7_deliver.REQ/ MM7_deliver.RES pair
Message type	Mandatory	Identifies this message as a MM7_deliver request
MM7 version	Mandatory	Identifies the version of the interface supported by the MMS relay/server
MMS relay/server ID	Optional	Identifier of the MMS relay/server
Linked ID	Optional	Identifier that may be used by the VASP in a subsequent MM7_submit.REQ
Sender address	Mandatory	The address of the MM originator
Recipient address	Optional	The address(es) of the intended recipients of the subsequent processing by the VASP or the original recipient address(es). It is possible to mark an address to be used only for informational purposes
Date and time	Optional	The time and date of the submission of the MM (time stamp)
Reply-charging ID	Optional	In case of reply-charging, when the reply MM is submitted within the MM7_ deliver.REQ, this is the identification of the original MM that is replied to
Priority	Optional	The priority (importance) of the message
Subject	Optional	The title of the whole MM
Content type	Mandatory	The content type of the MM
Content	Optional	The content of the MM

4.15.3.6 Distribution list adminstration

Once a VAS becomes available users may subscribe to the service using direct contact to the VASP (e.g., by sending an MM via MM1_submit.REQ to the service provider including registration information). The distribution list may be maintained by the

Table 4.5 Information elements in the MM7_deliver.RES

Information element	Presence	Description
Transaction ID	Mandatory	Identification of the MM7_deliver.REQ/ MM7_deliver.RES pair
Message type	Mandatory	Identifies this message as a MM7_deliver response
MM7 version	Mandatory	Identifies the version of the interface supported by the VASP
Service code	Optional	Information supplied by the VASP that may be included in charging information. The syntax and semantics of the content of this information are beyond the scope of this specification
Request status	Mandatory	Status of completion of the request
Request status text	Optional	Text description of the status for display purposes, qualifying the request status

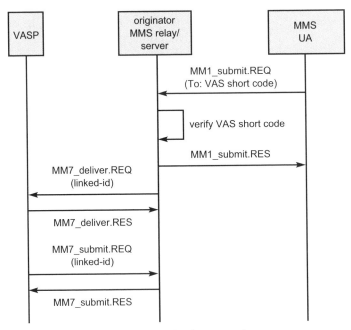

Figure 4.19 Use of MM7_deliver and subsequent response

MM7 = interface to VASP; VASP = Value Added Service Provider; MMS = Multimedia Messaging Service; UA = User Agent

MMS relay/server. The full definition of administration of the distribution list may be specified in future releases of this specification.

4.15.3.7 Implementation of MM7 abstract messages

The interface between a VASP and the MMS relay/server, over the MM7 reference point, is implemented using SOAP as the formatting language. The VASP and the MMS relay/server are able to play dual roles of sender and receiver of SOAP messages. HTTP is often used as the transport protocol of SOAP messages, which bind to the HTTP request/response model by providing SOAP request parameters in the body of the HTTP POST request and the SOAP response in the body of the corresponding HTTP response (Openwave, 2002).

4.15.3.8 SOAP message format and encoding principles

The following principles must be used in the design of SOAP implementation of the MM7 interface:

- The MM7 schema is based on W3C SOAP 1.1 and indicates the version of the MM7 specification that is supported. Note that the W3C SOAP 1.1 schema is published by 3GPP and that the Uniform Resource Identifier (URI) is **www.3gpp.org/ftp/Specs/ archive/23_series/23.140/schema/REL-5-MM7-1-1**

- MM7 SOAP messages consist of a SOAP envelope, SOAP header element and SOAP body element (see Figure 4.20).

- Transaction management is handled by the SOAP header element and transaction ID is included as a SOAP header entry. The SOAP "actor" attribute should not be specified in the SOAP header entry and The SOAP "mustUnderstand" attribute should be specified with value "1".

- All MM7 information elements, except for the transaction ID, should be included in the SOAP body element.

- XML element names use the Upper Camel Case convention, where words are concatenated to form an element name with

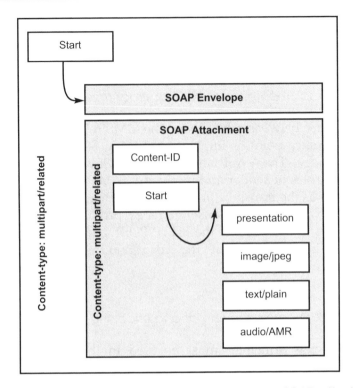

Figure 4.20 Message structure for a message with a SOAP attachment (multipart/ralted payload)

SOAP = Simple Object Access Protocol; AMR = Adaptive Multi Rate

the first letter of each word in upper case (e.g., EarliestDelivery-Time).

4.15.3.9 Binding to HTTP

MM7 request messages are transferred in an HTTP POST request. and responses are transferred in an HTTP response message. The content type "text/xml" is only used for messages containing the SOAP envelope. MM7 requests that carry a SOAP attachment have a "multipart/related" content type. The SOAP envelope is the first part of the MIME message and indicates the "Start" parameter of the multipart/related content type. If a SOAP attachment is included it is encoded as a MIME part and becomes the second part of the HTTP POST message. The MIME part should have the appropriate content type(s) to identify the payload. Figures 4.20

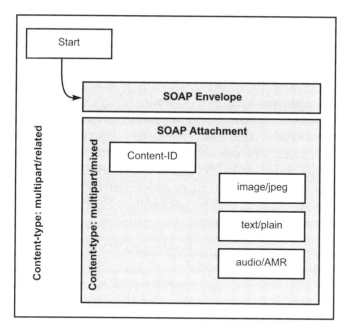

Figure 4.21 Message structure for a message with a SOAP attachment (multipart/mixed payload)

SOAP = Simple Object Access Protocol; AMR = Adaptive Multi Rate

and 4.21 provide examples of the message structure. This MIME part has two headers – content type and content ID (content ID is referenced by the MM7 request <Content> element).

4.15.3.10 MM7 addressing considerations

In order to bind properly to HTTP, the MMS relay/server and the VASP must be addressable by a unique URI-type address, which is placed in the host header field in the HTTP POST method (i.e., **vasp.host.com**). In the SOAP body, when the recipient MMS UA is addressed, the address-encoding scheme for MM1 is be used (i.e., +447710123456/TYPE=PLMN). The VASP is also identified by an MM1 address.

4.15.3.11 Status reporting

MM7 response messages are carried within an HTTP response, which may carry status at three levels:

- Network errors are indicated by the HTTP level (e.g., as an HTTP 403 "server not found") and are carried in the HTTP response back to the originating application.

- Request processing errors (status codes in the range 2xxx–9xxx) will be reported as a SOAP fault, which should include the "Faultcode", "Faultstring" and "Detail" elements. The "Detail" element includes the status elements described below and in Table 4.6. It also includes the "VASPErrorRsp" or "RSErrorRsp" element as direct child elements. The VASPErrorRsp element is included if the SOAP Fault is generated by the VASP and the RSErrorRsp element will be sent if the SOAP Fault is generated by the MMS relay/server.

- Also, errors relating to the transaction ID will be reported as a SOAP fault. The "faultcode" is of type "Client.TransactionID" and the "Faultstring" is used to indicate a human-readable "Description of the error." No "Detail" element is required.

Success or partial success (having a status code 1xxx from the "success" class) is reported in an MM7 response message that includes the following status elements, which are contained in the "Status" element of the response messages. All status responses reported are contained within three XML elements in the response (i.e., the details of the SOAP fault and the status of the MM7 response message). A status code indicates a numerical code that identifies different classes of error or successful completion of the operation. The status code is a four-digit number (the two high-order digits are defined on the page opposite and the two low-order digits are implementation-specific).

   Status text contains a predefined human-readable description of the numerical code that indicates the general type of error. The "Details" element is optional, but gives particular details of the error or partial success (e.g., indicates the address that cannot be resolved or message ID that is not recognized). The format of the "Details" element is implementation-specific.

4.15.3.12   Request and error status codes

The "StatusText" element (for application-level situations) is used to carry a human-readable explanation of the error or success

Table 4.6 ``StatusCode'' and ``StatusText''

StatusCode	StatusText	Meaning
1000	Success	Indicates that the request was executed completely
1100	Partial success	Indicates that the request was executed partially, but some parts of the request could not be completed. Lower order digits and the optional "details" element may indicate what parts of the request were not completed
2000	Client error	Client made an invalid request
2001	Operation restricted	The request was refused due to lack of permission to execute the command
2002	Address error	The address supplied in the request was not in a recognized format, or the MMS relay/server ascertained that the address was not valid for the network, because it was determined not to be serviced by this MMS relay/server. When used as a response result, with multiple recipients specified in the corresponding PUSH submission, this status code indicates that at least one address is incorrect
2003	Address not found	The address supplied in the request could not be located by the MMS relay/server. This code is returned when an operation is requested on a previously submitted message and the MMS relay/server cannot find the message for the address specified
2004	Multimedia content refused	The server could not parse the MIME content that was attached to the SOAP message and indicated by means of the "content" element that the content size or media type was unacceptable
2005	Message ID not found	This code is returned either when an operation is requested on a previously submitted message and the MMS relay/server cannot find the message for the message ID specified, or when the VASP receives a report concerning a previously

continued

Table 4.6 (cont.)

StatusCode	StatusText	Meaning
		submitted message and the message ID is not recognized
2006	Linked ID not found	This code is returned when a linked ID was supplied and the MMS relay/server could not find the related message
2007	Message format corrupt	An element value format is inappropriate or incorrect
3000	Server error	The server failed to fulfil an apparently valid request
3001	Not possible	The request could not be carried out because it is not possible. This code is normally used as a result of a cancel or status query on a message that is no longer available for cancel or status query. The MMS relay/server has recognized the message in question, but it cannot fulfil the request, because the message is already complete or status is no longer available
3002	Message rejected	Server could not complete the service requested
3003	Multiple addresses not supported	The MMS relay/server does not support this operation for multiple recipients. The operation *may* be resubmitted as multiple single-recipient operations
4000	General service error	The requested service cannot be fulfilled
4001	Improper identification	The ID header of the request does not uniquely identify the client (neither the VASP nor the MMS relay/server)
4002	Unsupported version	The version indicated by the MM7 "version" element is not supported
4003	Unsupported operation	The server does not support the request indicated by the "MessageType" element in the header of the message
4004	Validation error	The SOAP and XML structures could not be parsed, mandatory fields ar missing or the message format is not compatible with the format specified. The "details" field

StatusCode	StatusText	Meaning
		may specify the parsing error that caused this status
4005	Service error	The operation caused a server (either the MMS relay/server or the VASP) failure and should not be resent
4006	Service unavailable	Sent by the server when service is temporarily unavailable (e.g., when server is busy)
4007	Service denied	The client does not have permission or funds to perform the requested operation

situation (e.g., partial success). The status text should be used by the VASP or MMS relay/server when indicating status information to the originator. In addition to this there will be status codes consisting of a four-digit numeric value, the first digit of which indicates the class of the code. There are four classes:

1xxx   success in the operation;
2xxx   client errors;
3xxx   server errors;
4xxx   service errors.

Status codes are extensible. Both the VASP and the MMS relay/server must understand the class of a status code: unrecognized codes will be treated as the x000 code for that class, while codes outside the four defined class ranges will be treated as 3000. For implementation-specific codes, the numbers in the range x500–x999 should be used. Table 4.6 shows the status codes and status texts that are currently defined.

4.15.4   Mapping information elements to SOAP elements

The following subsections detail the mapping of information elements of abstract messages to SOAP elements. The full XML Schema definition of the MM7 reference point appears at

http://www.3gpp.org/ftp/Specs/archive/23-series/23.140/schema/REL-5-MM7-1-2
and the specification of the format that SOAP element values take
appears in the schema.

4.15.4.1 MM7_submit.REQ mapping

Information element	Location	ElementName	Comments
Transaction ID	SOAP header	TransactionID	
Message type	SOAP body	MessageType	Defined as "Root" element of SOAP body
MM7 version	SOAP body	MM7Version	Value is the number of this specification (e.g., 5.2.0)
VASP ID	SOAP body	VASPID	
VAS ID	SOAP body	VASID	
Sender address	SOAP body	SenderAddress	
Recipient address	SOAP body	Recipients	Different address format will be specified as part of element value
Service code	SOAP body	ServiceCode	Information supplied for billing purposes – exact format is implementation-dependent
Linked ID	SOAP body	LinkedID	Message ID of linked message
Message class	SOAP body	MessageClass	Enumeration – possible values: informational, advertisement, auto
Date and time	SOAP body	TimeStamp	
Time of expiry	SOAP body	ExpiryDate	
Earliest delivery time	SOAP body	EarliestDeliveryTime	
Delivery report	SOAP body	DeliveryReport	Boolean – true or false
Read reply	SOAP body	ReadReply	Boolean – true or false
Reply-charging	SOAP body	ReplyCharging	No value – presence implies "true"!

Reply deadline	SOAP body	ReplyDeadline	Attribute of "ReplyCharging" element's date format – absolute or relative
Reply-charging size	SOAP body	ReplyChargingSize	Attribute of "ReplyCharging" element
Priority	SOAP body	Priority	Enumeration – possible values: high, normal, low
Subject	SOAP body	Subject	
Adaptations	SOAP body	AllowAdaptations	Attribute of "Content" element (Boolean – true or false)
Charged party	SOAP body	ChargedParty	Enumeration – possible values: sender, recipient, both, neither
Message distribution indicator	SOAP body	DistributionIndicator	Boolean – true or false
Content type	MIME header – attachment	ContentType	
Content	SOAP body	Content	"href:cid" attribute links to attachment

4.15.4.2   MM7_submit.RES mapping

Information element	Location	ElementName	Comments
Transaction ID	SOAP header	Transaction ID	
Message type	SOAP body	MessageType	Defined as "Root" element of SOAP body
MM7 version	SOAP body	MM7Version	Value is the number of this specification (e.g., 5.2.0)
Message ID	SOAP body	MessageID	
Request status	SOAP body	StatusCode	See p. 147
Request status text	SOAP body	StatusText & Details	See p. 147

4.15.4.3   Sample message submission

```
POST /mms-rs/mm7 HTTP/1.1
Host: mms.omms.com
Content-Type: multipart/related;
boundary="NextPart_000_0028_01C19839.84698430"; type=text/xml;
  start="</tnn-200102/mm7-submit>"
Content-Length: nnnn
SOAPAction: ""
--NextPart_000_0028_01C19839.84698430
Content-Type:text/xml; charset="utf-8"
Content-ID: </tnn-200102/mm7-submit>

<?xml version="1.0" ?>
<env:Envelope xmlns:env="http://schemas.xmlsoap.org/soap/envelope/">
  <env:Header>
    <mm7:TransactionID
xmlns:mm7="http://www.3gpp.org/ftp/Specs/archive/23_series/23.140/schema/
REL-5-MM7-1-2" env:mustUnderstand="1">
     vas00001-sub
    </mm7:TransactionID>
  </env:Header>
  <env:Body>
    <SubmitReq
xmlns="http://www.3gpp.org/ftp/Specs/archive/23_series/23.140/schema/REL-
5-MM7-1-2">
      <MM7Version>5.3.0</MM7Version>
      <SenderIdentification>
        <VASPID>TNN</VASPID>
        <VASID>News</VASID>
      </SenderIdentification>
      <Recipients>
        <To>
          <Number>7255441234</Number>
          <RFC2822Address
<displayOnly="true">7255442222@OMMS.com</RFC2822Address>
        </To>
        <Cc>
          <Number>7255443333</Number>
        </Cx>
        <Bcc>
    <RFC2822Address>7255444444@OMMS.com</RFC2822Address>
        </Bcc>
      </Recipients>
      <ServiceCode>gold-sp33-im42</ServiceCode>
      <LinkedID>mms00016666</LinkedID>
      <MessageClass>Informational</MessageClass>
       <TimeStamp>2002-01-02T09:30:47-05:00</TimeStamp>
      <EarliestDeliveryTime>2002-01-02T09:30:47-
05:00</EarliestDeliveryTime>
      <ExpiryDate>P90D</ExpiryDate>
      <DeliveryReport>true</DeliveryReport>
      <Priority>Normal</Priority>
```

```
        <Subject>News for today</Subject>
        <ChargedParty>Sender</ChargedParty>
          <DistributionIndicator>true</DistributionIndicator>
          <Content href="cid:SaturnPics-01020930@news.tnn.com"
allowAdaptations="true"/>
        </SubmitReq>
    </env:Body>
</env:Envelope>

--NextPart_000_0028_01C19839.84698430
Content-Type: multipart/mixed; boundary="StoryParts 74526 8432 2002-77645"
Content-ID:<SaturnPics-01020930@news.tnn.com>

--Story 74526 8432 2002-77645
Content-Type: text/plain; charset="us-ascii"
Science news, new Saturn pictures ...

--StoryParts 74526 8432 2003-77645
Content-Type: image/gif;
Content-ID:<saturn.gif>
Content-Transfer-Encoding: base64
R0lGODdhZAAwAOMAAAAAAIGJjGltcDE0OOfWo6Ochbi1n1pmcbGojpKbnP/lpW54f
BMTE1RYXEFO
--StoryParts 74526 8432 2002-77645--
--NextPart_000_0028_01C19839.84698430--
```

Note that the different encoding mechanisms, as defined by
RFC2045, can be utilized for content encoding. The response
message is sent by the MMS relay/server back to the VASP for
the VAS application in an HTTP response message:

```
HTTP/1.1 200 OK
Content-Type: text/xml; charset="utf-8"
Content-Length: nnnn

<?xml version="1.0" ?>
<env:Envelope xmlns:env="http://schemas.xmlsoap.org/soap/envelope/">
  <env:Header>
    <mm7:TransactionID
xmlns:mm7="http://www.3gpp.org/ftp/Specs/archive/23_series/23.140/schema/
REL-5-MM7-1-2" env:mustUnderstand="1">
      vas00001-sub
    </mm7:TransactionID>
  </env:Header>
  <env:Body>
    <SubmitRsp
xmlns="http://www.3gpp.org/ftp/Specs/archive/23_series/23.140/schema/REL-
5-MM7-1-2">
      <MM7Version>5.3.0</MM7Version>
      <Status>
        <StatusCode>1000</StatusCode>
        <StatusText>Success</StatusText>
      </Status>
```

```
    <MessageID>041502073667</MessageID>
  </SubmitRsp>
 </env:Body>
</env:Envelope>
```

4.15.4.4   MM7_deliver.REQ mapping

Information element	Location	ElementName	Comments
Transaction ID	SOAP header	TransactionID	
Message type	SOAP body	MessageType	Defined as ''Root'' element of SOAP body
MM7 version	SOAP body	MM7Version	Value is the number of this specification (e.g., 5.2.0)
MMS relay/server ID	SOAP body	MMSRelayServerID	
Linked ID	SOAP body	LinkedID	Message ID of linked message
Sender address	SOAP body	Sender	
Recipient address	SOAP body	Recipients	If none appear, then the sender's address is used
Date and time	SOAP body	TimeStamp	
Reply-charging ID	SOAP body	ReplyCharging ID	Should correspond to an ID that appeared in a previous MM7_submit.REQ
Priority	SOAP body	Priority	Enumeration – possible values: high, normal, low
Subject	SOAP body	Subject	
Content type	MIME header of attachment	ContentType	
Content	SOAP body	Content	''href:cid'' attribute links to attachment

4.15.4.5 MM7_deliver.RES

Information element	Location	ElementName	Comments
Transaction ID	SOAP header	TransactionID	
Message type	SOAP body	MessageType	Defined as "Root" element of SOAP body
MM7 version	SOAP body	MM7Version	Value is number of this specification (e.g., 5.2.0)
Service code	SOAP body	ServiceCode	
Request status	SOAP body	StatusCode	See p. 147
Request status text	SOAP body	StatusText & Details	See p. 147

4.15.4.6 Sample deliver request and response

```
POST /mms/weather.xml HTTP/1.1
Host: www.yahoo.com
Content-Type:multipart/related;
boundary="NextPart_000_0125_01C19839.7237929064"; type=text/xml;
  start="</cmvt256/mm7-deliver>"
Content-Length: nnnn
SOAPAction: ""

--NextPart_000_0125_01C19839.7237929064
Content-Type:text/xml; charset="utf-8"
Content-ID: </cmvt256/mm7-submit>

<?xml version="1.0"?>
<env:Envelope xmlns:env="http://schemas.xmlsoap.org/soap/envelope/">
  <env:Header>
     <mm7:TransactionID
xmlns:mm7="http://www.3gpp.org/ftp/Specs/archive/23_series/23.140/schema/
REL-5-MM7-1-2" env:mustUnderstand="1">
      vas00324-dlvr
     </mm7:TransactionID>
   </env:Header>
   <env:Body>
     <!-- Example of MM7_deliverReq -->
     <DeliverReq
xmlns="http://www.3gpp.org/ftp/Specs/archive/23_series/23.140/schema/REL-
5-MM7-1-2">
      <MM7Version>5.3.0</MM7Version>
      <MMSRelayServerID>240.110.75.34</MMSRelayServerID>
      <LinkedID>wthr8391</LinkedID>
      <Sender>
```

```
    <RFC2822Address>97254265781@MMS.com</RFC2822Address>
        </Sender>
        <TimeStamp>2002-04-15T14:35:21-05:00</TimeStamp>
        <Priority>Normal</Priority>
        <Subject>Weather Forecast</Subject>
        <Content href="cid:forecast-location200102-86453"/>
        </DeliverReq>
    <env:Body>
</env:Envelope>

--NextPart_000_0125_01C19839.7237929064
Content-Type:text/plain;charset="utf-8"
Content-ID:<forecast-location2000102-86453>
Los Angeles, Calif, USA
--NextPart_000_0125_01C19839.7237929064--
```

The deliver response message might look like this (with an application error code):

```
HTTP/1.1 200 OK
Content-Type: text/xml; charset="utf-8"
Content-Length: nnnn

<?xml version="1.0"?>
<env:Envelope xmlns:env="http://schemas.xmlsoap.org/soap/envelope/">
  <env:Header>
      <mm7:TransactionID
xmlns:mm7="http://www.3gpp.org/ftp/Specs/archive/23_series/23.140/schema/
REL-5-MM7-1-2" env:mustUnderstand="1">
        vas00324-dlvr
      </mm7:TransactionID>
    </env:Header>
    <env:Body>
    <env:Fault>
        <faultcode>env:Client</faultcode>
        <faultstring>Client error</faultstring>
        <detail>
        <VASPErrorRsp
xmlns="http://www.3gpp.org/ftp/Specs/archive/23_series/23.140/schema/REL-
5-MM7-1-2">
            <MM7Version>5.3.0</MM7Version>
            <Status>
              <StatusCode>4006</StatusCode>
              <StatusText>Service
Unavailable</StatusText>
          <Details>Location not covered in service </Details>
            </Status>
          </ VASPErrorRsp>
        </detail>
      </env:Fault>
    </env:Body>
</env:Envelope>
```

4.15.5   MMS video streaming

This subsection analyses the current position and roadmap of video streaming in relation to the TS 23.140 MMS specification (3GPP, 2002a). It includes all key aspects involved in offering wireless video services, from handset and network video-related technologies and standards to mobile operators' plans to introduce mobile video services. Streaming refers to the ability of an application to play media streams, such as audio and video streams, in a continuous way to a client (or player) over a data network. Streaming not only reduces the requirements on available terminal memory and the need to reduce terminal costs but also enables the distribution of copyright media content files, such as live or premium services. The increasingly important issue of illegal copying of material is addressed, because the content is not saved on the terminal after the streaming session has ended. Content access, digital copyright management and streaming can be handled together, allowing a flexible and reliable way to distribute digital content on the mobile Internet (Northstream, 2002b).

4.15.6   Mobile video services

Currently, there are at least three different types of mobile video services: mobile video messaging, mobile video distribution services (including both streaming and download) and mobile video telephony:

- Mobile video messaging is a Person-to-Person (P2P) or a Person-to-Machine (P2M) communication service that consists of sending video content together with other media on a non-real time basis from mobile to mobile, mobile to PC or PC to mobile.

- Mobile video distribution services allow mobile users to either stream or download video content to their mobile devices. The main difference between download and streaming is that the latter allows transferred data to be processed as a steady and continuous stream by a streaming multimedia application and displayed before the entire file has been transmitted (see Chapter

7, which deals with the Multimedia Broadcast/Multicast Service [MBMS]).

- Video telephony is a P2P communication service that enables a real-time, two-way stream of video and audio signals between two mobile devices or a mobile device and a fixed videophone.

Services that can be delivered using video streaming as a basis can be classified into on-demand and live information delivery applications. Examples of the first category are music and news-on-demand applications. Live delivery of radio and television programmes are examples of the second category. It is important to understand the characteristics of each one of these mobile video services, as they impose different technical requirements both on the terminals and mobile networks. Although some of these services will be technically feasible following the introduction of GPRS, the more demanding mobile video services (such as the quality of service [QoS] requirements imposed by mobile video telephony) will only be a reality when they are based on dedicated bandwidth resources (as offered by Third Generation [3G] or Universal Mobile Telecommunications System [UMTS] networks).

4.15.7 Terminal support for video services

3GPP-compliant, MPEG-4 or H.263-based (www.m4if.org) video telephones will constitute the next generation of video-enabled phones. It is still unclear when these terminals will reach global critical mass (these videophones are today a reality in Japan and parts of Europe). Even though the range of mobile terminals will become much wider than today, not all future mobile terminals will be video-enabled. As we described on p. 18, the terminal replacement rate and the penetration of data-related devices are key factors for understanding mobile video services uptake. A slower acceptance of video-enabled devices would have a significant impact on the success and uptake of all mobile video services.

4.15.8 Network support for video services

The transition from Two and a Half Generation (2.5G) to 3G mobile networks will enable mobile applications that demand higher data rates and more stringent QoS requirements. With the widespread introduction (in Europe and the USA) of GPRS technology, mobile video messaging services and downloading of small video files into the mobile device are now possible, due to the increased data rates provided and the non real time characteristics of these services.

Streamed video distribution services are best supported by 3G networks. However, niche video applications could have a place on GPRS networks offering small, low-resolution video files that require low data throughput. Streaming full-length movies to a phone or handheld computer will be an impractical application for a long time.

Some mobile network technologies are not appropriate for real-time video telephony/conferencing. For instance, GPRS latency (approx >250ms), packet loss (>1%), low data rate (<30kbps) and lack of QoS characteristics make it an unsuitable technology for real-time video telephony. 3G is the first technology that can be widely used for video telephony services, as it offers circuit-switched data rates of 64kbps both to and from the terminal. In addition to GPRS and 3G mobile networks, there are other wireless, broadband technologies capable of high data rates that will also enable mobile video services within selective areas. These include both WLAN and Wireless Personal Area Network (WPAN) technologies, such as 802.11b and Bluetooth.

4.15.9 Packet Streaming Service

The PSS provides a framework for IP-based streaming applications in mobile networks (Elsen et al., 2001; 3GPP, 2002e). The release of the 3GPP MMS specification allows delivery of messages that consist of slide show presentations containing still images, text and audio. MMS enables rich content to be created for messages while also functioning as a highly versatile platform for many mobile applications and services. MMs can also be sent, stored and forwarded to and from Internet applications. The messages will be pushed to the receiving terminal or email account in a way that is similar to the SMS.

Eventually, video clips will be part of MM content. To embed a video clip into MMS content would limit the video to approximately 5–15 seconds (90KB). Therefore, when media content becomes more demanding in terms of message size, streaming

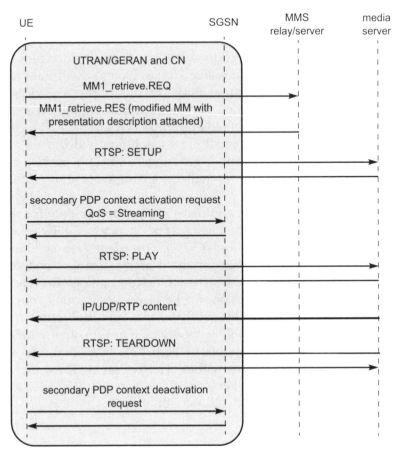

Figure 4.22 Schematic view of the support for streaming in MMS

MMS = Multimedia Messaging Service; UTRAN = Universal Terrestrial Radio Network; GERAN = GSM/EDGE Radio Network; GSM = Global System for Mobile; EDGE = Enhanced Data rate for GSM Extension; CN = Core Network; MM1 = interface between user agent and MMS relay/server; RTSP = Real Time Streaming Protocol; QoS = Quality of Service; IP = Internet Protocol; UDP = User Datagram Protocol; RTP = Real-time Transport Protocol; PDP = Packet Data Protocol; SGSN = Servicing GPRS Support Node; GPRS = General Packet Radio Service

will become necessary in order to deliver the messages from the server to the receiving terminal or Internet application. This functionality has been planned for inclusion in a future release of the MMS specification.

Figure 4.22 shows an example of the transaction flow for using streaming in MMS for retrieval of streamable MM elements. The MMS relay/server sends a modified MM as MM1_retrieve.RES in response to a retrieve request (MM1_retrieve.REQ). Note that the

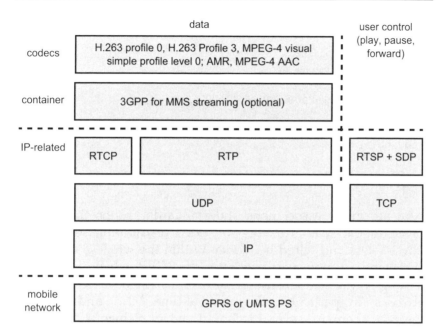

Figure 4.23 3GPP video streaming stack
3GPP = Third Generation Partnership Project; AMR = Adaptive Multi Rate; MMS = Multimedia Messaging Service; RTCP = Real-time Control Protocol; RTP = Real-time Transport Protocol; RTSP = Real Time Streaming Protocol; SDP = Session Description Protocol; UDP = User Datagram Protocol; TCP = Transmission Control Protocol; IP = Internet Protocol; GPRS = General Packet Radio Service; UMTS = Universal Mobile Telecommunications System; PS = Packet Streaming

interface between the MMS relay/server and the media server is not specified in release 5.

Session Description Protocol (SDP) is used as the format of the presentation description. The attribute line ("a=") with "Control" type in the SDP header indicates the need to open a Real Time Streaming Protocol (RTSP) session. The use of RTSP is to set up and control a streaming session (Figure 4.23 shows a video streaming stack). An example of a presentation description in SDP format, describing the streaming of a video sequence, now follows:

```
v=0
o=ghost 2890844526 2890842807 IN IP4 192.168.10.10
s=MMS Example
i=Example of SDP file for streaming in MMS
u=http://www.mediaserver.com/ae600
e=ghost@server.com
c=IN IP4 0.0.0.0
```

```
b=AS:128
t=0 0
a=range:npt=0-45.678
m=video 1024 RTP/AVP 96
b=AS:128
a=rtpmap:96 H263-2000/90000
a=fmtp:96 profile=3;level=10
a=control:rtsp://test.titaniumnet.ws/realmedia/test/test.rm
a=recvonly
```

4.15.10 Video standards

There are many associations and standards groups that are involved in the standardization of video distribution services for
both wireless and wired networks. Within the wireless world, the
main standard-setting association is the 3GPP, which is currently
working on the standardization of wireless video services over 3G
networks (in particular, the non-real-time MMS and the PSS.
However, there are several technical and implementation aspects
that have been left for future standardization work. These include
other implementation options for MMS, security mechanisms and
DRM for PSS.

   Over the last two years, several industry fora have been created
with the purpose of ensuring and promoting end-to-end interoperability and helping the development of different aspects of
the mobile video market. Some of the most relevant ones include
the Wireless Media Forum (WMF), the MPEG-4 Industry Forum
(M4IF), the Internet Streaming Media Alliance (ISMA) and the
International Telecommunications Union (ITU) H.263 codec.
MPEG and ITU standards development is under way on the
H.264/MPEG4 part 10 codec, which is expected to offer significant
improvement in bandwidth and quality in the next few years.

4.15.11 Video coding

One of the key issues is the use of a commonly accepted video
coding standard to enable universal compatibility between terminals and services. Over the last few years the Internet has experienced a fierce "video format war" with multiple proprietary video
compression standards fighting for the highest market share. This
phenomenon continues in the wireless world, with most mobile
terminals supporting one or several of the RealVideo, Windows

Media and MPEG4, H.263 and H.264 formats. It remains to be seen whether the MPEG4 or the joint H.264/MPEG4 variant of the video standard will become the dominant format. Although MPEG4 has been seen as the natural succesor to MPEG2, there is significant hardware development of H.264 codecs by major manufacturers. VideoLocus is a leading designer of reference hardware for the video coding market (**www.videolocus.com**).

PDAs and other mobile devices with increased computational capacity may either integrate or allow downloading from other proprietary Internet video players like Windows Media Player, RealNetwork or QuickTime. It should be noted that these players also support MPEG4 codecs and other (proprietary) codec mechanisms specific to the player developer.

There are numerous companies developing MPEG4 products, from software-based players to device chipsets and media servers. Most of these MPEG4 products will include features that offer enhancements to the standard, as these companies will aim to differentiate themselves from the competition. Therefore, it is also crucial to agree on the minimum subset of video codec functionality to allow effective implementation of a standard within mobile devices. The current approach by Nokia is for their mobile devices to support both H.263 and MPEG4 video codecs and to do so uses the Real video player (Nokia, 2002b).

4.15.12 Open source video tools

The Helix community is a collaborative effort between RealNetworks, independent developers and leading companies to create and extend the Helix DNA platform, the first open platform for digital media delivery. The Helix DNA platform comprises the following:

- Helix DNA client;
- Helix DNA producer;
- Helix DNA server.

The Helix community is made up of many constituencies, each with different goals in addition to a shared objective of continuing to extend the Helix platform. To allow developers and their organizations the maximum amount of flexibility, RealNetworks has elected to utilize multiple licenses, both open and commercial, each tailored to a different constituency and need. The Helix DNA client is available for a range of OSs including Pocket PC and Symbian.

4.15.13   The challenge for mobile operators

The key issues for the operator concerning mobile video service deployment are the service roadmap, pricing and user education. The mobile operator must take advantage of the opportunity offered by video as a means to increase airtime traffic, revenue, service differentiation and customer loyalty. It should be noted though that services with video content will initially be low volume compared with other text and voice-based services. The mobile operator must define a service roadmap including realistic video applications, but should prioritize those services that are expected to generate more revenues in the short term. Finding viable pricing models for mobile video services is a challenge, as high charges may hinder the general usage of mobile video applications. Additionally, the operator should take into account that mobile subscribers need to be educated and become familiar with these new services.

Other implementation and technical issues that mobile operators must take into consideration when delivering video are:

- content management (also including content creation/acquisition and content hosting);

- content adaptation and/or capability exchange (depending on terminal capability);

- mechanisms to dynamically and automatically adjust the data rate;

- traffic dimensioning in order to support applications with higher bandwidth requirements;

- DRM.

4.16 CONCLUDING REMARKS

In this chapter we have taken a detailed look at the applications layer and the many technical issues facing VASPs and developers of MMS content services. We have suggested practical solutions to some of these issues, from application development and testing to deployment and subsequent content management. The determination of terminal capability using CC/PP has been outlined, as have been the development tools for building SMIL/MMS content and the compilation and distribution of MMS files. This includes a description of the MMS WAP-PUSH notification used to inform a user of the arrival of an MM. The topic of integrating video streaming to MMS devices also helps identify further milestones on the roadmap for service and application evolution.

5

Network Layer

5.1 INTRODUCTION

This chapter will detail the network technologies required in a
Multimedia Messaging Service (MMS) system. By network we
mean the core systems at the heart of MMS. This does not
include the handset technology, or the applications, although
both will be touched on (already detailed in Chapter 4). The core
systems are the Wireless Application Protocol (WAP) gateway (the
component required to move the content), the persistent storage
(used to store content in the network), the MMS server/relay that
controls movement of the messages and notification (as well as
interfacing with foreign systems) and the billing system with its
associated mechanisms. The chapter will also include a discussion
on an approach toward implementation of network components,
covering open source software, and will look at the availability of
existing products. The issue of interoperability is covered as it is
considered to be extremely important to the success of MMS: if
subscribers cannot exchange messages between networks and
accounting systems between operators are not in place, then it is
guaranteed to be a failure.

A few incumbent suppliers have traditionally guarded telecommunications networks. The voice network's vital organs have
been exchanges, billing systems and Operations Support Systems
(OSSs). The arteries and nerves of these networks have been
transmission systems and protocols, which have been unique to

Multimedia Messaging Service Daniel Ralph and Paul Graham
© 2004 John Wiley & Sons, Ltd ISBN: 0-470-86116-9

telecommunications (e.g., the signalling system number 7 suite of protocols). Even though these protocols are standardized, suppliers and operators have adopted their own proprietary variations to prevent outside companies (both operators and equipment suppliers) getting in.

MMS changes the model described above. It is based on standardized protocols, with all possible interfaces well defined, permitting operators and suppliers not normally associated with telecommunications to get into the area. The cost barriers to entry have been significantly reduced. The standard for MMS has adopted technologies that are familiar to a wide audience of developers, including Extensible Mark-up Language (XML), Simple Mail Transfer Protocol (SMTP) and Common Object Request Broking Architecture (CORBA). The standards body has also ensured that all information required for billing is available through the path of a message: Charging Data Records (CDRs) are generated at many trigger points, so enabling novel tariffing schemes to be devised. The details of these CDRs have been included in this chapter with a view to inspiring the reader to come up with his or her own tariffing scheme.

5.2   MMS NETWORK ELEMENTS

This section outlines and specifies the network elements and interfaces required to implement an MMS system for an operator or service provider. An MMS network contains the following elements:

- MMS relay/server;

- WAP gateway;

- Short Message Service Centre (SMSC);

- MMS client;

- billing system;

- persistent storage.

We will now discuss the network elements required to construct a complete MMS system and permit it to interwork with external systems.

5.2.1 MMS relay/server

The MMS relay/server is responsible for storage and notification, reports and general handling of messages. It may also provide convergence functionality between external servers and MMS User Agents (UAs) and thus enable the integration of different server types across different networks. It is possible to separate the MMS relay/server element into MMS relay and MMS Server elements, but allocation of the MMS Relay/Server functionalities to such elements is not currently defined. The MMS relay/server provides the following functionalities:

- receiving and sending Multimedia Messages (MMs);

- conversion of messages arriving at the recipient MMS relay/ server from legacy messaging systems to MM format (e.g., facsimile to MM) if interworking with legacy messaging systems (MM3) is supported;

- conversion of MMs leaving the originator MMS relay/server to legacy messaging systems to the appropriate message format (e.g., MM to Internet email) if interworking with legacy messaging systems (MM3) is supported;

- message content retrieval;

- MM notification to the MMS UA;

- generating delivery reports;

- routing forward MMs and read–reply reports;

- address translation;

- temporary storage of messages;

- ensuring that messages are not lost until successfully delivered to another MMS Environment (MMSE) element.

The MMS relay/server should provide such additional functionalities as:

- generating CDRs;
- negotiation of terminal capabilities.

The MMS relay/server may provide such additional functionalities as:

- MM forwarding;
- address hiding;
- persistent storage of messages;
- controlling the reply-charging feature of MMS;
- relaying message distribution indicators.

The MMS relay/server is able to provide such additional features as:

- enabling/disabling the MMS function;
- personalizing the MMS, based on user profile information;
- MM deletion, based on user profile or filtering information;
- media type conversion;
- media format conversion;
- screening of MM;
- checking terminal availability;
- managing the message properties on servers (e.g., voicemail or email server) integrated in the MMSE (consistency) (only applicable if interworking with legacy messaging systems [MM3] is supported).

This list of additional optional functionalities of the MMS relay/server is not exhaustive.

5.3 PERSISTENT NETWORK-BASED STORAGE

An optional feature of MMS is the support of persistent, network-based storage called an "MMBox" (Multimedia Message Box), a logical entity associated with the MMS relay/server where MMs may be stored, retrieved and deleted. Depending on an operator's configuration, each subscriber may have his or her MMBox configured to automatically store incoming and submitted MMs, or, by means of the supporting MMS UAs, request that specific MMs be persistently stored on a case-by-case basis.

5.3.1 MMS user databases and Home Location Register (HLR)

The MMS may have access to several user databases, which may consist, for example, of a user profile database, subscription database and HLR. Such user databases provide:

- MMS user subscription information;

- information for the control of access to the MMS;

- information for the control of the extent of available service capability (e.g., server storage space);

- a set of rules on how to handle incoming messages and their delivery;

- information about the current capabilities of the user's terminal.

5.3.2 Billing system

All Multimedia Messaging Service Centres (MMSCs) are required to be intergrated with a billing system for charging and accounting purposes. Operaters and service providers usually have legacy systems that they are obliged to use.

5.3.3 External servers

Several external servers may be included within or connected to an MMSE (e.g., email server, Short Message Service [SMS] server [SMSC], fax). Convergence functionality between external servers and MMS UAs is provided by the MMS relay/server, which enables the integration of different server types across different networks.

5.4   MMSC NETWORK INTERFACES

5.4.1   MM3: MMS relay/server–external servers

Reference point MM3 is used by the MMS relay/server to send MMs to and retrieve them from servers of external (legacy) messaging systems that are connected to the service provider's MMS relay/server, such as SMSC elements.

5.4.2   MM4: interworking of different MMSEs

Reference point MM4 between MMS relay/servers belonging to different MMSEs is used to transfer messages between them.

5.4.3   MM5: MMS relay/server–HLR

Reference point MM5 may be used to provide information to the MMS relay/server about the subscriber. If this reference point is provided, then it uses existing Mobile Application Part (MAP) operations, which are part of Signalling System No. 7 (SS7) (e.g., procedures for determining the location of the mobile, or for alerting SMSs). When SMS is used as the bearer for notification, this reference point is not necessary.

5.4.4   MM6: MMS relay/server–MMS
user databases

This interface provides interworking between external user databases, such as those holding user profile information (currently, this is undefined).

5.4.5   MM8: MMS relay/server–billing system

This interface provides interworking with an external billing system. Currently, as of release 5 of the Third Generation Partnership Project (3GPP) specifications, this is undefined.

5.5 WAP GATEWAY

The method used to transport the content in the MMS system application is defined under the WAP standard. The WAP protocol is an industry-wide specification for developing applications that operate over wireless communications networks. A WAP gateway is a piece of software that has several functions in the "chain" between the WAP device and the Web server. These functions are in general:

- converting the mark-up language (WML [Wireless Mark-up Language]/XHTML [Extensible Hyper Text Mark-up Language]) from textual format to tokenized (a binary/compressed) format that is readable by the WAP device (WAP browser);

- translating the requests from the WAP device to Hyper Text Transfer Protocol (HTTP) requests for the Web world;

- converting between the SSL (Secure Sockets Layer) encryption used in the Web world and the Wireless Transport Layer Security (WTLS) encryption used in the WAP world;

- converting between the "transport" protocol of the Web (TCP [Transmission Control Protocol]) and that of the WAP world (WDP [Wireless Datagram Protocol]).

Optionally, many gateways also perform other conversions, such as between plain text files or simple HTML files into WAP-readable format. If a mobile operator hosts a WAP gateway, the mobile operator may use the gateway to add mobile-specific information into the HTTP stream, such as the subscriber number and location details (although this is not widely done). Some WAP gateways are publicly available on the Internet, while others are hosted by mobile operators. The latter are usually restricted to accept WAP devices coming from the mobile operator's dial-in service only.

The growth of MMS services is placing heavy demands on conventional WAP gateways, which are not ready to deal with the heavy bursts of activity that can be associated with bulk MMS delivery. The increased burden on WAP gateways will require optimizations to be able to handle the increased load. Support for Segmentation And Reassembly (SAR) will be essential if the gateway is to support larger object delivery to MMS clients. SAR

is supported to provide a mechanism for larger object delivery to
MMS clients.

5.5.1 Wireless Transaction Protocol – Segmentation and Reassembly

As new mobile phones with colour displays, increased memory,
multimedia and high-speed packet communication capabilities
are being developed, mobile operators and service providers are
increasingly dependent on efficient server solutions in their infrastructures. There is an increasing need for improved data speeds
and overall service quality, both of which are restrained by a
narrow air interface. In line with these requirements the WAP
Forum defines, in addition to compressed data format (WBXML
[WAP Binary XML]), a lightweight protocol with several enhancements that reduce the number of bytes sent through the air interface. Some of the enhancements are mandatory, whereas others
may be deployed as optional features. One such optional enhancement is the Wireless Transaction Protocol Segmentation And Reassembly (WTP SAR) feature (Figure 5.1). This feature is essential to
enable downloading of MMs of up to 30KB in size.

5.6 LARGE-SCALE MAIL SYSTEMS

The market success and growth of the SMS in Europe shows that
the MMS has the potential to be one of the most lucrative
mobile services in the future. As we have seen, MMS will allow
mobile subscribers to exchange not only simple text but also high-
resolution pictures, audio, animations and video to and from a
mobile phone. For MMS to become a quick success, operators
must effectively address certain key areas, such as persistent
content storage, to ensure that Person-to-Person (P2P) traffic
escalates at a fast pace from day 1. The communication potential
of P2P messaging will require the ability to store MMs in an album
for future retrieval. This is aligned with the view that P2P messaging is not just an application but a critical success factor of the
service. These advanced features are conceived and designed with
the aim of helping to generate a potent "viral effect" in MMS,

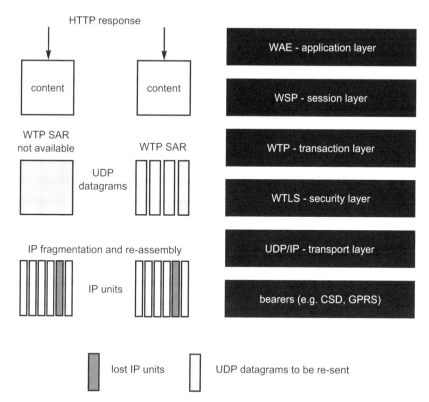

Figure 5.1 Location of WTP SAR in the WAP protocol stack

WTP SAR = Wireless Transaction Protocol Segmentation and Reassembly; WAP = Wireless Application Protocol; HTTP = Hyper Text Transfer Protocol; UDP = User Datagram Protocol; IP = Internet Protocol; WAE = Wireless Application Environment; WSP = WAP Session Protocol; WTP = Wireless Transaction Protocol; WTLS = Wireless Transport Layer Security; CSD = Circuit Switched Data; GPRS = General Packet Radio Service

meaning making it easy for end-users to get hold of MMS content, to pass it on to other users and to respond to incoming messages in an easy and convenient way.

In the 3GPP architecture the server will be an email server providing post office services that are accessible (e.g., via POP3 [Post Office Protocol version 3] or IMAP) for Internet email retrieval in the MMSE or are accessible to the MMS relay/server using SMTP. The MMS relay/server will send messages that are to be transmitted as Internet email via SMTP. In the case of retrieval and sending of MMs from and to an Internet email service (Figure 5.2), this too is done via SMTP, the protocol used on the interface between MMS relay/server and the mail transfer agent.

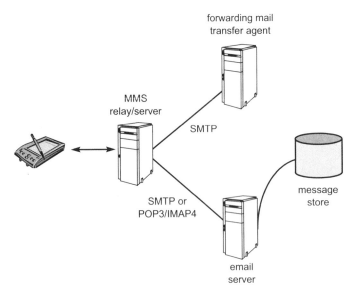

Figure 5.2 Example of interaction with Internet email messaging
SMTP = Simple Mail Transfer Protocol; MMS = Multimedia Messaging Service; POP3 = Post Office
Protocol version 3; IMAP4 = Internet Message Access Protocol version 4

5.6.1 Persistent network-based storage

As with the SMS the core-enabling technology utilized for MMS is a
store-and-forward platform that pushes the messages to users once
the recipient's handset becomes available. Does both the pro-
visional or transient character and the limited size of text messages
mean there is no necessity for a place to store them perpetually
outside the users' terminals? On the contrary, an MM is typically
much larger and has higher emotional value than a 160-character
note, due to its enhanced multimedia content. In addition, an MM
is more complex to create. Users will want to store them for longer
periods of time, probably for several months (Frost and Sullivan,
2002; Patel, 2002). To illustrate this, assume a hypothetical number
of messages handled by a subscriber per day (e.g., between 1 and
10 messages). Next, assume how long they wish to store them (e.g.,
between one week and three months). This means there would be
between 7 and 900 messages that would need storage in the
network, 225 as the average. We should take into consideration
that if the handset can store an average of 20 MMs this figure
would actually be lower. A network-based persistent store or per-

manent store will prove to be an indispensable element for the MMS for additional reasons to those already mentioned:

- The handset has limited memory for storage. Should users run out of storage space on the device, they could keep messages in the MMS persistent store for the long term.

- Users without a persistent store who are deprived of or mislay their phones will lose all their personal multimedia content.

- Owners of terminals with a poor display will be able to render their MMs from a PC at another time and enjoy greater resolution (in the case of picture messaging).

- Those subscribers who do not wish to receive messages while roaming can store them temporarily and receive them later (or, again, via a different type of accessing device).

- Users do not want to get rid of or delete premium MMS content they have paid for.

- Having or increasing storage on a handset adds costs to every subscriber – whether they use it or not. Offering network-based persistent storage allows handset manufacturers to minimize the cost of storage on devices.

But, most importantly, it is self-fulfilling that users who can store and possess an unlimited amount of MMs will send out more messages. This facility is especially relevant to early adopters, so they can ignite greater interest in MMS usage. This service also offers the possibility to charge for persistent storage. Subscribers could be given a free storage space (say, 1MB), then offered the chance to purchase further amounts using an SMS message (i.e, "You have exceeded your 1Meg quota, reply if you would like to purchase another meg").

It should be noted that many devices are beginning to support memory card slots (such as Memory Stick duo) and the MMS format. These cards can be transferred between a PC and phone and so may obviate some of the need for online storage.

5.6.2 Multimedia Message Box (MMBox)

MMBox is the specification for network-based storage for MMS. It allows all users to have their own MMBox, a place to store and retrieve MMS messages.

Table 5.1 Abstract messages for storing or updating stored MMs

Abstract messages	Type	Direction
MM1_mmbox_store.REQ	Request	MMS UA → MMS relay/server
MM1_mmbox_store.RES	Response	MMS UA ← MMS relay/server

5.6.3 Storing and updating MMs in an MMBox

Requests from an MMS UA to store MMs will always be sent to the corresponding MMS relay/server. The abstract messages involved are outlined in Table 5.1 from type and direction points of view.

5.6.3.1 Normal operation

The MMS UA submits a request to store an MM in the MMBox using the MM1_mmbox_store.REQ, which contains the message reference received in the MM1_notification.REQ. In addition, the MMS UA submits a request to update the MM state and/or MM flags of an MM already stored within an MMBox using the MM1_mmbox_store.REQ, which contains the message reference, MM state and/or MM flags obtained from any previous operation resulting in an MM being stored or updated in the MMBox. The MMS relay/server responds with an MM1_mmbox_store.RES, which provides the status of the store or MM update request. The MM1_mmbox_store.RES unambiguously refers to the corresponding MM1_mmbox_store.REQ. Support for MM1_mmbox_ store transactions are optional for the MMS UA and mandatory for the MMS relay/server, if MMBoxes are supported.

5.6.3.2 Abnormal operation

In this case the MMS relay/server responds with a MM1_mmbox_ store.RES that gives the current status and indicates the reason the MM was not able to be stored or updated (e.g., service not available, MMBoxes not supported, MMBox not enabled, MMBox over quota, MMBox system full, MMBox system I/O error). If the MMS relay/server does not provide the MM1_mmbox_store.RES, the MMS UA should assume that the MM was not stored or updated, and should be able to recover it.

5.6.3.3 Features

- *Message Reference* The Message Reference in MM1_mmbox_store.REQ indicates the MM to be stored or updated. This reference can be from MM1_notification.REQ, or the message reference from any of the store request responses (e.g., MM1_mmbox_store.RES, MM1_mmbox_view.RES, MM1_forward.RES with Store, MM1_submit.RES with Store). The message reference in MM1_mmbox_store.RES indicates a reference to the newly stored or updated MM that is helpful for subsequent usage.

- *MM State* The MMS UA may request that the MM be either stored, or updated, with a specific MM State. In the absence of this value, when the Message Reference refers to a new MM (i.e., from MM1_notification.REQ) the default will be the new state. In the absence of this value, when the Message Reference refers to an MM already stored the MM State will not be changed.

- *MM Flags* If present, there are one or more keyword values. In the absence of this element, no values are assumed for newly stored MMs and no changes made for already stored MMs.

- *Store Status* The MMS relay/server indicates the status of the MM1_mmbox_store.REQ in the Store Status information element of the associated MM1_mmbox_store.RES. The Store Status information element of the MM1_mmbox_store.RES may be supported with explanatory text. If this text is available in the Store Status Text information element the MMS UA should bring it to the user's attention. The choice of the language used in the Store Status Text information element is at the discretion of the MMS service provider.

- *Transaction Identification* The MMS UA provides unambiguous transaction identification within a request. The response will unambiguously refer to the corresponding request using the same transaction identification.

- *Version* The MMS protocol willl provide unique means to identify the current version of the particular protocol environment.

- *Message Type* The type of the message used on the reference point MM1, indicating MM1_mmbox_store.REQ and MM1_mmbox_store.RES as such.

5.6.3.4 *Information elements in the MM1_mmbox_store.REQ*

Information element	Description
Message Type	Identifies this message as MM1_mmbox_store.REQ
Transaction ID	Identifies the MM1_mmbox_store.REQ/MM1_mmbox_store.RES pair
MMS Version	Identifies the version of the interface supported by the MMS UA
Message Reference	Message reference from a MM1_notification.REQ or any previous store or MMBox view operation
MM State	State of the MM: if absent when the message reference is from a notification request it defaults to "New". No value is assumed when the message reference refers to an already stored MM
MM Flags	Keyword flags of the MM. There are no defaults

5.6.3.5 *Information elements in the MM1_mmbox_store.RES*

Information element	Description
Message Type	Identifies this message as MM1_mmbox_store.RES
Transaction ID	Identifies the MM1_mmbox_store.REQ/MM1_mmbox_store.RES pair
MMS Version	Identifies the version of the interface supported by the MMS relay/server
Message reference	Reference to the newly stored or updated MM that is helpful for subsequent usage (e.g., with MM1_retrieve.REQ and MM1_mmbox_delete.REQ)
Store status	Status of the MM store operation
Store Status Text	Description that qualifies the status of the MM store request

Table 5.2 Abstract messages for viewing the MMBox

Abstract messages	Type	Direction
MM1_mmbox_view.REQ	Request	MMS UA → MMS relay/server
MM1_mmbox_view.RES	Response	MMS UA ← MMS relay/server

5.6.3.6 View the MMBox

This part of the MMS describes the mechanism by which an MMS UA may request a listing of the MMs contained within the subscriber's MMBox. The MMS UA issues the request to view selected portions of MMs within the subscriber's MMBox, as well as information about the MMBox itself, from the corresponding MMS relay/server. The abstract messages involved are outlined in Table 5.2 from type and direction points of view.

5.6.3.7 Normal operations

The MMS UA will issue an MM1_mmbox_view.REQ message, containing optional request qualifiers, to the MMS relay/server. The MMS relay/server will respond with an abstract message, MM1_mmbox_view.RES, containing the resulting view data as the content of the abstract message. This information consists of a listing of the MMBox contents, possibly including information about the MMBox itself. When the Start and Limit attributes are used, several pairs of MM1 mmbox_view.REQ and MM1_mmbox_view.RES transactions might be used in order to acquire the complete set of results.

5.6.3.8 Abnormal operations

In this case the originator MMS relay/server will respond with a MM1_mmbox_view.RES that gives the current status and indicates the reason the operation could not be completed (e.g., corrupted abstract message, no subscription, service not available, MMBox not supported, MMBox not enabled, MMBox I/O error). If the

MMS relay/server does not provide the MM1_mmbox_view.RES, the MMS UA should be able to recover the MM.

5.6.3.9   Features

- *Attributes List*   A list of information element names used in the MM1_mmbox_view.REQ that request corresponding information elements from the MMs to be conveyed in the MM1_mmbox_view.RES. The list of known information element names are those currently defined for the MM1_retrieve.RES and MM1_notification.REQ. In the absence of the Attributes List information element, the MMS relay/server, by default and if available, selects the following information elements from each viewed MM: Message ID, Date and time, Sender address, Subject, Message size, MM State and MM Flags.

- *Message Selection*   Messages that are to be viewed may be selected by a list of Message References or by selection based on the MM State and/or MM Flags keywords. Either Message Reference List or Select may be supplied in the MM1_mmbox_view.REQ, which selects MMs for inclusion in the content of the MM1_mmbox_view.RES. In the absence of the Message Reference List, if Select is present and if any of the select keywords matches either the MM State or any of the MM flags of an MM in the MMBox, the requested information elements of the MM will be included in the MM1_mmbox_view.RES (e.g., "Select: new" or "Select: draft"). The absence of both the Message References List and the Select information elements will yield a listing of all MMs currently stored within the MMBox.

- *Partial Views*   MMBox View results may be received in their entirety or may be indexed to start the view at a given MM offset relative to the selected MMs, and/or may be limited to a finite number of MMs. The Start information element is a number that may be used in the MM1_mmbox_view.REQ to index the first MM to be viewed, relative to the selected set of MMs, allowing partial views to be requested. If Start is absent, the first selected MM will begin the view results. The Limit information element is a number that may be provided in the MM1_mmbox_view.REQ to specify a limit to the number of MMs with information elements that will be returned in the

MM1_mmbox_view.RES. If Limit is absent, all of the remaining MMs are returned.

- *MMBox Information* The Totals information element, if present in the request, indicates that the MMBox totals are requested. In the response the Totals information element value will be the total number of messages and/or total size, with the units (e.g., messages or bytes) identified. The Quotas information element, if present in the request, indicates that the MMBox quotas, in terms of messages and/or size, are requested. In the response the Quotas information element value refers to the quotas as the maximum number of messages allowed and/or the maximum size allowed, with the units (e.g., messages or bytes) identified.

- *MM Listing* A list of information elements from the MMs returned within the MM1_mmbox_view.RES. The listing consists of the following information elements, separately grouped for each MM returned in the list: Message reference (a unique reference to an MM) and Information elements corresponding to those requested in the Select information element in the MM1_mmbox_view.REQ.

- *Request Status* This will be the status code for any failures of the MM1_mmbox_view.REQ command. The "reason" code given in the status information element of the MM1_mmbox_view.RES may be supported with explanatory text, further qualifying the status. If this text is available in the Request status text information element the MMS UA should bring it to the user's attention. The choice of the language used in the Request status text information element is at the discretion of the MMS service provider.

- *Transaction Identification* The MMS UA provides an unambiguous transaction identification within a request. The response will unambiguously refer to the corresponding request, using the same transaction identification.

- *Version* The MMS protocol shall provide unique means to identify the current version of the particular protocol environment.

- *Message Type* The type of message used on the reference point MM1, indicating MM1_mmbox_view.REQ and MM1_mmbox_view.RES in this case.

5.6.3.10   Information elements in the MM1_mmbox_view.REQ

Information element	Description
Message Type	Identifies this message as MM1_mmbox_view.REQ
Transaction ID	Identifies the MM1_mmbox_view.REQ/MM1_mmbox_view.RES pair
MMS Version	Identifies the version of the interface supported by the MMS UA
Attributes List	List of information elements that are to be returned as a group for each MM listed in the MM1_mmbox_view.RES. If absent, the default list will apply
Message Reference List	One or more Message References that are to have their information elements listed
Select	List of MM State or MM Flags keywords, by which MMs within the MMBox can be selected, if the Message Reference list is absent
Start	Number indicating the index of the first MM of those selected to have information elements returned in the response. If this is absent, the first item selected is returned
Limit	Number indicating the maximum number of selected MMs to have their information elements returned in the response. If this is absent, information elements from all remaining MMs are returned
Totals	Indicates that the current total number of messages in the MMBox and/or the byte size of the MMBox are requested
Quotas	Indicates that the current message and/or size quotas are requested

5.6.3.11   Information elements in the MM1_mmbox_view.RES

Information element	Description
Message Type	Identifies this message as MM1_mmbox_view.RES
Transaction ID	Identifies the MM1_mmbox_view.REQ/MM1_mmbox_view.RES pair

MMS Version	Identifies the version of the interface supported by the MMS relay/server
MM Listing	The requested listing of the selected MMs, which will be one or more groups of information elements, one for each MM listed. Each MM group will include a Message Reference and possibly additional information elements as well. If absent, no MMs were found or selected
Request Status	If an error occurs, this is the code that indicates the exact cause. For successful responses the Status may be returned with a corresponding success code.
Request Status Text	If an error occurs, this may contain explanatory text that corresponds to the Request Status
Totals	Total number of messages in the MMBox and/or byte size of the MMBox, depending on the presence of Totals in the request
Quotas	Quotas of the MMBox measured in number of messages and/or byte size of the MMBox, depending on the presence of Quotas in the request

5.6.3.12 *Service records for a MMS relay/server that supports MMBoxes*

5.6.3.12.1 MMBox MM1 Store CDR (Bx1S-CDR)

If enabled, an MMBox MM1 Store CDR (Bx1S-CDR) will be produced in the MMS relay/server as long as the MMS relay/server responds with an MM1_mmbox_store.RES to the MMS UA:

Field	Description
Record Type	MMBox MM1 Store record
MMS Relay/Server Address	Address of the MMS relay/server
Managing Address	Address of the managing MMS US (i.e., the MMS UA that has sent the MM1_mmbox_store.REQ)
Access Correlation	Unique identifier delivered by the used access network domain of the originator MMS UA
Content Type	Content type of the MM content
Message Size	Size of the MM (bytes)

continued

Field	Description
Message Reference	Reference to the newly stored or updated MM that is helpful for subsequent usage (e.g., with MM1_retrieve. REQ and MM1_mmbox_delete.REQ)
MM State	State of the MM. If absent when the Message Reference is from a notification request it defaults to "New". No value is assumed when the Message Reference refers to an already stored MM
MM Flags	If available, the keyword flags of the MM. There are no defaults
Store Status	Status code of the request to store the MM as received in the MM1_store.RES
Store Status Text	This field includes a more detailed technical description of the store status precisely when the CDR is generated. This field is only present if the store status is present
Sequence Number	Record number
Time Stamp	Time of generation of the CDR
Record Extensions	Set of network/manufacturer-specific extensions to the record

5.6.3.12.2 MMBox MM1 View CDR (Bx1V-CDR)

If enabled, an MMBox MM1 View CDR (Bx1V-CDR) will be produced in the MMS relay/server as long as the MMS relay/server has sent an MM1_mmbox_view.RES to the MMS UA:

Field	Description
Record Type	MMBox MM1 View record
MMS Relay/Server Address	Address of the MMS relay/server
Managing Address	Address of the managing MMS UA (i.e., the MMS UA that has sent the MM1_mmbox_view.REQ)
Access Correlation	Unique identifier delivered by the used access network domain of the originator MMS UA
Attributes List	List of information elements that are to be returned as a group for each MM listed in the MM1_mmbox_view.RES. If absent, the default list (i.e., Message ID, Date and time,

	Sender address, Subject, Message size, MM State and MM Flags) will apply
Message Selection	List of MM State or MM Flag keywords (e.g., new or draft) or a list of Message References by which MMs within the MMBox can be selected. If both are absent, a listing of all MMs currently stored within the MMBox shall be selected
Start	Number indicating the index of the first MM of those selected to have information elements returned in the response. If this is absent, the first item selected is returned
Limit	Number indicating the maximum number of selected MMs to have their information elements returned in the response. If this is absent, information elements from all remaining MMs are returned
Totals Requested	This field indicates whether the current total number of messages in the MMBox and/or the byte size of the MMBox has been requested by the managing MMS UA
Quotas Requested	This field indicates whether the current message and/or size of quotas (i.e., the maximum number of messages allowed and/or the maximum size allowed) has been requested by the managing MMS UA
MM Listing	Requested listing of the selected MMs, which will be one or more groups of information elements, one for each MM listed. Each MM group will include a Message Reference and possibly additional information elements as well. If absent, no MMs were found or selected
Request Status Code	Status code of the request to view the MM as received in the MM1_view.RES
Status Text	This field includes the status text as received in the MM1_view.RES that corresponds to the Request Status Code. Present only if provided in the MM1_view.RES
Totals	Total number of messages and/or octets in the MMBox, depending on the presence of Totals in the request
Quotas	Quotas of the MMBox in messages and/or octets, depending on the presence of Quotas in the request
Sequence Number	Record number
Time Stamp	Time of generation of the CDR
Record Extensions	A set of network/manufacturer-specific extensions to the record

5.7 ACCESS TO MMBOXES USING LEGACY GSM HANDSETS

Using such an access technology as WAP or Unstructured Supplementary Services Data (USSD), it would be possible to allow users with legacy Global System for Mobile communications (GSM) handsets to use network-based persistent stores. Although they could not view the content using their handsets, it would be possible to allow them to store content and send it using a simple menuing system. Received content could be stored in the persistent store and accessed across the Internet using a PC. This system would be another source of revenue for operators and another method of getting Average Revenue Per User (ARPU). The problem with this though is that it may prevent users from moving across to MMS handsets, so the service and tariffing should be such that it not only gives users a taste of MMS but also encourages them to take on the full service and the handset or service agreement required.

5.8 MMBOX PEER-TO-PEER SHARING

Peer-to-peer file services are now very common in the Internet world. This feature could be applied to MMBoxes, giving users the opportunity to share part of their MMBoxes and allowing users to download messages or move them into their own area. This could be a compelling service: a piece of content could be described by word of mouth and then downloaded in much the same way as somebody would be told about a piece of music and then retrieve it using one of the many file sharing services. This would be a significant method of creating revenue as long as the necessary Digital Rights Management (DRM) systems are in place.

5.9 MOBILE DATA BILLING SYSTEMS AND THE MM8 INTERFACE

Charging for MMS presents operators with some difficult choices. MMS uses more network bandwidth than SMS, so one school of thought is that the user can be charged much more to send it.

However, high MMS prices will deter senders from using multi-media content, and, in any case, the price currently charged by operators for sending an SMS is typically many times higher than the cost of carrying it. On the other hand, from the marketing perspective, simple uniform pricing for MMS would be preferred, though this would fail to account for the wide variation in network resource consumption by messages of different sizes. Moreover, many different approaches will be investigated to ascertain the optimum way of charging for messages sent to and from legacy terminals and email systems.

It will be some time before the situation regarding MMS charging settles down and prevalent models begin to emerge. MMSCs must therefore support a flexible, experimental approach to charging, enabling an operator to bill according to a wide variety of events and parameters. The most obvious method of charging is by using CDRs, which can be collected by a billing system and processed. CDR parameters that might be supported include:

- MMS-specific message ID, recipient address(es) and sender address;
- message size (sent/received);
- identification (when a message has been sent to a predefined group);
- time stamp (including time zone) for submission time, earliest delivery time and time of expiry;
- duration of transmission;
- duration of storage in the MMS server;
- type of message;
- bearer type used;
- content information;
- message class;
- delivery report request;
- read–reply request;
- charging indicator (e.g., prepaid, reply, reversed or third-party financed);

- status (e.g., delivered, abandoned, time expired or delivery pending);
- indication of forwarding.

Also suggested for future consideration:

- specific class for instant messaging functionality;
- conversion of type and media;
- security level;
- priority/quality of service.

Figure 5.3 illustrates the Third Generation (3G) charging logical architecture, which is subdivided into two transmission planes: the Circuit Switched (CS) domain and the Packet Switched (PS) domain. The CDRs generated by the serving nodes (SGSN [Sensing GPRS Support Node], GGSN [Gateway GPRS Support Node]) for the appropriate domain are forwarded via the Charging Gateway Function (CGF) entities to the billing system for processing. Note that the Service Capability Features (SCFs) may also transfer CDRs directly to the billing system. However, the current specifications do not include any CDR descriptions for the SCF. (While not shown explicitly in this figure, the VLR may also generate CDRs.) CDRs for the MMS are delivered by the MMS relay/server when receiving or delivering MMs to the MMS UA or to another MMSE. CDRs from the MMS relay/server are transferred directly to the billing system. Such applications/services such MMS and LCS are provided to 3G subscribers via service nodes. These servers (service nodes) are responsible for the provision of application services to a subscriber and can generate a service-related CDR to record the details of the service transaction provided.

5.9.1 Charging information

Charging information for use by the MMS is collected for each MS by the MMS relay/server that serves the MMS UA. The information that the operator uses to raise a subscriber invoice is operator-specific. Note that billing aspects (e.g., a regular fee for a fixed period) are beyond the scope of the present document. The

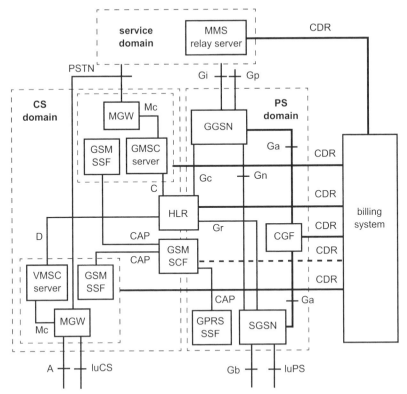

Figure 5.3 3GPP charging architecture

3GPP = Third Generation Partnership Project; MMS = Multimedia Messaging Service; PSTN = Public Switched Telephone Network; CS = Circuit Switched; PS = Packet Switched; CDR = Charge Detail Record; MGW = media gateway; GSM = Global System for Mobile communications; SSF = Service Switching Function; GMSC = Gateway Mobile Switching Centre; GGSN = Gateway GPRS Support Node; HLR = Home Location Register; CGF = Charging Gateway Function; SCF = Service Capability Feature; VMSC = Visited Mobile Switching Centre; CAP = CAMEL Application Part; CAMEL = Customized Applications for Mobile Network Enhanced Logic; GPRS = General Packet Radio Service; SGSN = Serving GPRS Support Node. Source: 3GPP

MMS relay/server collects charging information for each MS about any value-added service provided and usage of any MMS-specific network resources. The MMS relay/server will collect the following charging information:

- usage of MMS resources: the charging information describes the amount of data transmitted in both MO (Mobile Originating) and MT (Mobile Terminating) directions for the transfer of MMs;

- storage duration: MM storage duration is counted as either the time interval from the beginning of storage of the message until forwarding to another MMS relay/server or as the time interval from the beginning of storage until reception of the MM by an MMS UA (the time interval an MM is saved in non-volatile memory media);

- usage of general PS domain resources: the charging information will describe the usage of other PS domain-related resources;
- destination and source: the charging information will provide the actual destination and source addresses used by the subscriber;
- usage of external data networks: the charging information will describe the amount of data sent to and received from external data networks;
- the MMS relay/server address: this provides the highest accuracy location information available.

5.9.1.1 Charging scenarios

Let us now look at some sample scenarios that illustrate the purpose and practical usage of the various types of records defined in Table 5.3. The events triggering the generation of

Table 5.3 Record type overview for combined MMS relay/server

Trigger point	Trigger name
1	Originator MM1 submission
2	Recipient MM1 notification request
3	Recipient MM1 notification response
4	Recipient MM1 retrieval
5	Recipient MM1 acknowledgement
6	Originator MM1 delivery report
7	Recipient MM1 read–reply recipient
8	Originator MM4 read–reply originator
Any time between 1 ... 8*	Originator MM deletion

* No CDR will be generated as a result of receiving MM1 UA-initiated transactions (i.e., MM1_submit. REQ and MM1_retrieve.REQ)

Table 5.4 Trigger-type overview for the originator MMS relay/server

Trigger point	Trigger name
A1	Originator MM1 submission
A2	Originator MM4 forward request
A3	Originator MM4 forward response
A4	Originator MM4 delivery report
A5	Originator MM1 delivery report
A6	Originator MM4 read–reply report
A7	Originator MM1 read–reply originator
Any time between A1 … A7	Originator MM deletion

CDRs (Table 5.4) take place at the MM1 reference point and/or at the MM4 reference point (3GPP, 2002b). The MMS relay/servers generate CDRs when receiving MMs from or when delivering MMs to the UA or another MMS relay/server (Figures 5.4–5.6). The label in the message flows identifies the CDR generation trigger. The events triggering the generation of CDRs take place at the MM1 reference point and/or at the MM4 reference point.

5.9.1.2   Record description

Dedicated types of CDRs can be generated in the service domain for MMS by MMS relay/servers. For each CDR type the field definition includes the field name, description and category. The MMS relay/server will be able to provide CDRs at the billing system interface in the format and encoding described directly below. Additional CDR formats and contents, generated by the MMS relay/server, may be available at the interface to the billing system to meet the requirements of the billing system.

5.9.1.3   MMS records for originator MMS relay/server

The following subsections specify examples of CDRs created in the originator MMS relay/server, based on messages flowing over the MM1 and MM4 reference points (3GPP, 2002b). CDRs that refer to

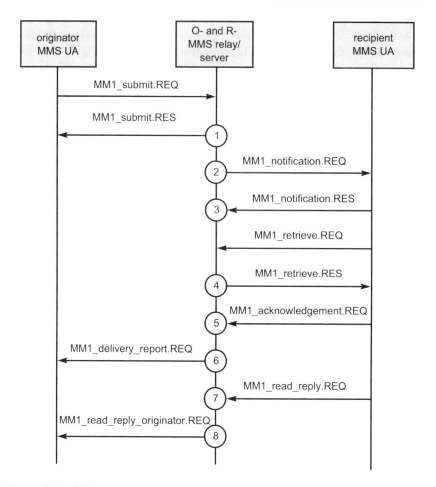

Figure 5.4 Originator and recipient MMS relay/server are the same
MMS UA = Multimedia Messaging Service User Agent; O = Originator; R = Recipient. Source: 3GPP (2002b)

MM4 messages (Originator MM4 *** CDR) are created only if the originator and recipient MMS relay/servers communicate over the MM4 interface (i.e., the originator MMS relay/server is not also the recipient MMS relay/server). The CDRs referring to MM1 messages (Originator MM1 *** CDR) are created regardless of whether the originator MMS relay/server is also the recipient MMS relay/server or not. Unless otherwise specified, the CDR parameters are copied from the corresponding MM1 or MM4 message parameters, as applicable.

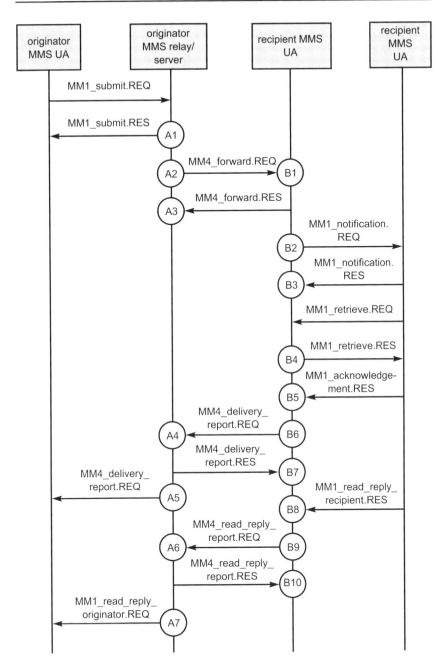

Figure 5.5 Originator and recipient MMS relay/server are not the same

MMS UA = Multimedia Messaging Service User Agent. Source: 3GPP (2002b)

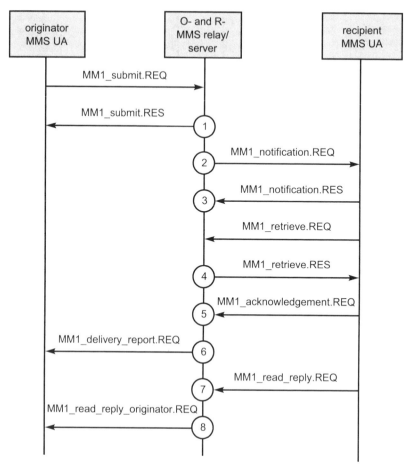

Figure 5.6 Originator and recipient MMS relay/server are the same

MMS UA = MultiMedia Messaging Service User Agent; O = Originator; R = Recipient; MM1 = MultiMedia 1. Source: 3GPP (2002b)

5.9.1.3.1 Originator MM1 Submission CDR

If enabled, an Originator MM1 Submission Charging Data Record (O1S-CDR) (Table 5.5) is produced in the originator MMS relay/ server for each MM submitted in an MM1_submit.REQ by an originator MMS UA to the originator MMS relay/server as long as the originator MMS relay/server responds with an MM1_submit.RES. The operator can configure the system so that this CDR, if enabled, will be created only for MM1_submit.RES (indicating acceptance of the submitted MM), or for unsuccessful submissions as well. Note that this includes the case where the MM is a reply to

Table 5.5 Originator MM1 Submission CDR (O1S-CDR)

Field	Description
Record Type	Originator MM1 submission record
Originator MMS Relay/Server Address	IP address or domain name of originator MMS relay/ server
Message ID	MM identification provided by the originator MMS relay/server
Reply-Charging ID	This field is present in the CDR only if the MM is a reply to an original MM. The reply-charging ID is the message ID of the original MM
Originator Address	Address of the originator MMS UA (i.e., the MMS UA that has sent the MM1_submit.REQ)
Recipients Address List	Address(es) of the recipient MMS UA(s) of the MM. Multiple addresses are possible if the MM is not a reply
Access Correlation	Unique identifier delivered by the used access network domain of the originator MMS UA
Content Type	The content type of the MM
MM Component List	List of media components with volume size
Message Size	Total size of the MM content (bytes)
Message Class	Class selection (i.e., personal, advertisement, information service) if specified in the MM1_submit. REQ
Charge Information	Charge indication and type
Submission Time	Time at which the MM was submitted from the originator MMS UA if specified in the MM1_submit. REQ
Time of Expiry	Desired date of expiry, or duration of time prior to expiry, for the MM if specified by the originator MMS UA
Earliest Time of Delivery	This field contains either the earliest time to deliver the MM or the number of seconds to wait before delivering the MM (as specified by the originator MMS UA)
Duration of Transmission	Time used for transmission of the MM between the UA Agent and the MMS relay/server

continued

Field	Description
Request Status Code	Status code of the MM as received in the MM1_submit. REQ
Delivery Report Requested	This field indicates whether a delivery report has been requested by the originator MMS UA or not
Reply Charging	Request for reply-charging, if specified by the originator MMS UA
Reply Deadline	In case of reply-charging, the latest time of submission of replies granted to the recipient(s) (as specified by the originator MMS UA)
Reply Charging Size	In case of reply-charging, the maximum size for replies granted to the recipient(s) (as specified by the originator MMS UA)
Priority	The priority (importance) of the message, if specified by the originator MMS UA
Sender Visibility	A request to show or hide the sender's identity when the message is delivered to the recipient (as specified by the originator MMS UA)
Read–Reply Requested	Request for read–reply report (as specified in the MM1_submit.REQ)
Status Text	This field includes a more detailed technical status of the message precisely when the CDR is generated (only present if the MM submission is rejected)
Record Time Stamp	Time of generation of the CDR
Local Record Sequence Number	Consecutive record number created by this node (the number is allocated sequentially and covers all CDR types)
MMBox Storage Information	Set of parameters related to MMBox management. This parameter is only present if the MMBox feature is supported by the MMS relay/server and storage of the MM was requested by the originator MMS UA (i.e., the MMS UA that sent the MM1_submit.REQ)
Record Extensions	Set of network/manufacturer-specific extensions to the record (conditional on the existence of an extension)

an original MM. In this case the MMS UA sending the reply is called the originator MMS UA of the reply and the MMS relay/server receiving the reply in an MM1_submit.REQ is called the originator MMS relay/server for the reply.

5.9.1.3.2 Originator MM4 Forward Request CDR (O4FRq-CDR)

If enabled, an Originator MM4 Forward Request Charging Data Record (O4FRq-CDR) (Table 5.6) will be produced in the originator MMS relay/server as long as the originator MMS relay/server has sent an MM4_forward.REQ to the recipient MMS relay/server, regardless of whether or not a MM4_forward.RES is received from the recipient (i.e., the CDR is created on completion of transmission of the MM4_forward.REQ). The MM4_forward.REQ may be generated as a reaction to an incoming MM1_forward.REQ. In this case, the "Originator address" field specifies the address of the originator MMS UA of the original MM, whereas the address of the forwarding MMS UA is contained in the "Forwarding address" field.

5.9.1.3.3 Originator MM4 Delivery report CDR (O4D-CDR)

If enabled, an Originator MM4 Delivery report CDR (O4D-CDR) (Table 5.7) will be produced in the originator MMS relay/server as long as the originator MMS relay/server receives an MM4_delivery_report.REQ from the recipient MMS relay/server. This CDR can be used by a billing system to correlate with the originally sent CDR to permit charging to continue.

5.9.1.3.4 Originator MM1 Delivery report CDR (O1D-CDR)

If enabled, an Originator MM1 Delivery report CDR (O1D-CDR) will be produced in the originator MMS relay/server as long as the originator MMS relay/server sends an MM1_delivery_report.REQ to the originator MMS UA.

5.9.1.3.5 Originator MM4 Read–reply report CDR (O4R-CDR)

If enabled, an Originator MM4 Read–reply report CDR (O4R-CDR) (Table 5.8) will be produced in the originator MMS relay/server as long as the originator MMS Relay/Server receives an MM4_read_reply_report.REQ from the recipient MMS relay/server. This CDR can be used to indicate to the orginal sender that an MMS message has been read, just as a read–reply can be generated by an email system.

5.9.1.3.6 Originator MM1 Read–reply originator CDR (O1R-CDR)

If enabled, an Originator MM1 Read–reply originator CDR (O1R-CDR) will be produced in the originator MMS relay/server as long

Table 5.6 Originator MM4 Forward Request CDR (O4FRq-CDR)

Field	Description
Record Type	Originator MM4 forward request record
Originator MMS Relay/Server Address	IP address or domain name of the originator MMS relay/server
Message ID	MM identification provided by the originator MMS relay/server
3GPP MMS Version	MMS version of the originator MMS relay/server
Originator Address	Address of the originator MMS UA of the MM. If the MM4_forward.REQ is generated as a reaction to an incoming MM1_forward.REQ, this is the address of the originator MMS UA of the original MM
Recipients Address List	Address(es) of the recipient MMS UA(s) of the MM (as specified in the MM4_forward.REQ that triggered the CDR)
Content Type	Content type of the MM
MM Component List	List of media components with volume size
Message Size	Total size of MM content (bytes)
Message Class	Class of the MM (e.g., personal, advertisement, information service) if specified by the originator MMS UA
Submission Time	Time at which the MM was submitted or forwarded as specified in the corresponding MM1_submit.REQ or MM1_forwarding.REQ
Time of Expiry	Desired date of expiry, or duration of time prior to expiry, for the MM if specified by the originator MMS UA
Delivery Report Requested	This field indicates whether a delivery report has been requested by the originator MMS UA or not
Priority	Priority (importance) of the message, if specified by the originator MMS UA
Sender Visibility	Request to show or hide the sender's identity when the message is delivered to the MM recipient, if the originator MMS UA has requested his or her address to be hidden from the recipient
Read–Reply Requested	Request for read–reply report, if the originator MMS UA has requested a read–reply report for the MM
Acknowledgement Request	Request for MM4_forward.RES
Forward Counter	Counter indicating the number of times the particular MM was forwarded

Field	Description
Forwarding Address	Address(es) of the forwarding MMS UA(s). Multiple addresses are possible. In the multiple address case, this is a sequential list of the address(es) of the forwarding MMS UAs that forwarded the same MM
Record Time Stamp	Time of generation of the CDR
Local Record Sequence Number	Consecutive record number created by this node (the number is allocated sequentially and covers all CDR types)
Record Extensions	Set of network/manufacturer-specific extensions to the record (conditional on the existence of an extension)

Table 5.7 Originator MM4 Delivery report CDR (O4D-CDR)

Record Type	Originator MM4 delivery report record
Recipient MMS Relay/Server Address	IP address or domain name of the recipient MMS relay/server
Originator MMS Relay/Server Address	IP address or domain name of the originator MMS relay/server
Message ID	MM identification provided by the originator MMS relay/server
3GPP MMS Version	MMS version of the recipient MMS relay/server
Originator Address	Address of the originator MMS UA of the MM
Recipient Address	Address of the MM recipient
MM Date and Time	Date and time the MM was handled (retrieved, expired, rejected, etc.) (as specified in the MM4_delivery_report)
Acknowledgement Request	Request for MM4_delivery_report.RES
MM Status Code	Status code of the delivered MM (as received in the MM4_delivery_report.REQ)
Status Text	This field includes the status text (as received in the MM4_delivery_report.REQ) that corresponds to the MM status code (only present if provided in the MM4_delivery_report.REQ)
Record Time Stamp	Time of generation of the CDR
Local Record Sequence Number	Consecutive record number created by this node (the number is allocated sequentially and covers all CDR types)
Record Extensions	Set of network/manufacturer-specific extensions to the record (conditional on the existence of an extension)

Table 5.8 Originator MM4 Read–reply report CDR (O4R-CDR)

Field	Description
Record Type	Originator MM4 read–reply report record
Recipient MMS Relay/Server Address	IP address or domain name of the recipient MMS relay/server
Originator MMS Relay/Server Address	IP address or domain name of the originator MMS relay/server
Message ID	MM identification provided by the originator MMS relay/server
3GPP MMS Version	MMS version of the recipient MMS relay/server
Originator Address	Address of the originator MMS UA of the MM
Recipient Address	Address of the MM recipient
MM Date and Time	Date and time the MM was handled (retrieved, expired, rejected, etc.)
Acknowledgement Request	Request for MM4_read_reply_report.RES
Read Status	Status of the MM (as received in the MM4_read_reply_report.REQ)
Status Text	This field includes the status text (as received in the MM4_read_reply_report.REQ) that corresponds to the read status (only present if provided in the MM4_read_reply_report.REQ)
Record Time Stamp	Time of generation of the CDR
Local Record Sequence Number	Consecutive record number created by this node (the number is allocated sequentially and covers all CDR types)
Record Extensions	Set of network/manufacturer-specific extensions to the record (conditional on the existence of an extension)

as the originator MMS relay/server sends an MM1_read_reply_originator.REQ to the originator MMS UA.

5.9.1.3.7 Originator MM Deletion CDR (OMD-CDR)

If enabled, an Originator MM Deletion CDR (OMD-CDR) will be produced in the originator MMS relay/server after sending an MM1_submit.RES to the originator MMS UA as long as:

- the originator MMS relay/server decides to abandon processing the MM at any point after receiving the corresponding MM1_submit.REQ; or

- the originator MMS relay/server decides to delete the MM because of expiry of storage time, which may either be indicated in the submit request or governed by operator procedure (e.g., after successful MM delivery).

When the processing of an MM is abandoned or the MM is deleted, there remains no trace of it in the originator MMS relay/server. This CDR is created regardless of whether the originator MMS relay/server is also the recipient MMS relay/server or not. The usefulness of this is that, when an MMS is removed completely from the MMSC, this CDR is automatically generated; thus a record of the MMS is retained.

5.9.1.4 MMS records for recipient MMS relay/server

The following subsections specify CDRs created in the recipient MMS relay/server based on messages flowing over the MM1 and MM4 interfaces. The CDRs referring to MM4 messages (Recipient MM4 *** CDR) are only created if the originator and recipient MMS relay servers communicate over the MM4 interface (i.e., the recipient MMS relay/server is not also the originator MMS relay/server). CDRs that refer to MM1 messages (Recipient MM1 *** CDR) are created regardless of whether the recipient MMS relay/server is also the originator MMS relay/server or not. Unless otherwise specified, CDR parameters are copied from the corresponding MM1 or MM4 message parameters as applicable. This set of CDRs is important for MMs sent between different providers not only for accounting purposes but for many other uses as well.

5.9.1.4.1 Recipient MM4 Forward CDR (R4F-CDR)
If enabled, a Recipient MM4 Forward CDR (R4F-CDR) will be produced in the recipient MMS relay/server as long as the recipient MMS relay/server receives an MM4_forward.REQ from the originator MMS relay/server.

5.9.1.4.2 Recipient MM1 Notification Request CDR (R1NRq-CDR)
If enabled, a Recipient MM1 Notification Request CDR (R1NRq-CDR) will be produced in the recipient MMS relay/server as long

as the recipient MMS Relay/Server sends an MM1_notification. REQ to the recipient MMS UA. The CDR, besides containing the usual auditing information, also contains details about reply-charging and reply-charging size (preventing large and costly messages being sent).

5.9.1.4.3 Recipient MM1 Notification Response CDR (R1Ns-CDR)

If enabled, a Recipient MM1 Notification Response CDR (R1NRs-CDR) will be produced in the recipient MMS relay/server as long as the recipient MMS relay/server receives an MM1_notification.RES from the recipient MMS UA.

5.9.1.4.4 Recipient MM1 Retrieve CDR (R1Rt-CDR)

If enabled, a Recipient MM1 Retrieve Response CDR (R1Rt-CDR) will be produced in the recipient MMS relay/server as long as the recipient MMS Relay/Server has sent a MM1_retrieve.RES to the recipient MMS UA (i.e., the CDR is created on completion of transmission of the MM1_retrieve.RES).

5.9.1.4.5 Acknowledgement CDR (R1A-CDR)

If enabled, a Recipient MM1 Acknowledgement CDR (R1A-CDR) will be produced in the recipient MMS relay/server as long as the recipient MMS relay/server receives an MM1_acknowledgement. REQ from the recipient MMS UA.

5.9.1.4.6 Recipient MM4 Delivery report Request CDR (R4DRq-CDR)

If enabled, a Recipient MM4 Delivery report Request CDR (R4DRq-CDR) (Table 5.9) will be produced in the recipient MMS relay/server as long as the recipient MMS relay/server sends an MM4_delivery_report.REQ to the originator MMS relay/server.

5.9.1.4.7 Recipient MM4 Delivery report Response CDR (R4DRs-CDR)

If enabled, a Recipient MM4 Delivery report Response CDR (R4DRs-CDR) will be produced in the recipient MMS relay/server as long as the recipient MMS relay/server receives an MM4_delivery_report.RES from the originator MMS relay/server. The main purpose of this CDR would be to correlate it with the original CDR, to prevent the subscriber being charged for an MMS that was never delivered or to avoid unnecessary accounting taking place between operators.

Table 5.9 Recipient MM4 Delivery report Request CDR (R4DRq-CDR)

Field	Description
Record Type	Recipient MM4 delivery report request record
Recipient MMS Relay/Server Address	IP address or domain name of the recipient MMS relay/server
Originator MMS Relay/Server Address	IP address or domain name of the originator MMS relay/server
Message ID	MM identification provided by the originator MMS relay/server
3GPP MMS Version	MMS version of the recipient MMS relay/server
Originator Address	Address of the originator MMS UA of the MM
Recipient Address	Address of the MM recipient
MM Date and Time	Date and time the MM was handled (retrieved, expired, rejected, etc.)
Acknowledgement Request	Request for MM4_delivery_report.RES
MM Status Code	Status code of the MM (as sent in MM4_delivery_report.REQ)
Status Text	This field includes the status text (as sent in the MM4_delivery_report.REQ) that corresponds to the MM status code
Record Time Stamp	Time of generation of the CDR
Local Record Sequence Number	Consecutive record number created by this node (the number is allocated sequentially and covers all CDR types)
Record Extensions	Set of network/manufacturer-specific extensions to the record (conditional on the existence of an extension)

5.9.1.4.8 Recipient MM4 Read–reply report Request CDR (R4RRq-CDR)

If enabled, a Recipient MM4 Read–reply report Request CDR (R4RRq-CDR) will be produced in the recipient MMS relay/server as long as the recipient MMS relay/server sends an MM4_read_reply_report.REQ to the originator MMS relay/

server. This can be used to notify the sender that the MM sent was
read by the recipient.

5.9.1.4.9   Recipient MM4 Read–reply report Response CDR
(R4RRs-CDR)

If enabled, a Recipient MM4 Read–reply report Response CDR
(R4RRs-CDR) will be produced in the recipient MMS relay/
server as long as the recipient MMS relay/server receives an
MM4_read_reply_report.RES from the originator MMS relay/
server.

5.9.1.4.10   Recipient MM Deletion CDR (RMD-CDR)

If enabled, a Recipient MM Deletion CDR (RMD-CDR) will be
produced in the recipient MMS relay/server as long as:

- the recipient MMS Relay/Server decides to abandon processing
  of the MM at any point after receiving the corresponding
  MM4_forward.REQ; or

- the recipient MMS Relay/Server decides to delete the MM
  because of expiry of storage time, which may either be indicated
  in the submit request or governed by operator procedure (e.g.,
  after successful MM delivery).

When the processing of the MM is abandoned, there remains no
trace of the MM in the recipient MMS Relay/Server. The status
code indicates the precise reason for abandoning or deleting the
MM. A special case occurs when the recipient MMS relay/server
is also the forwarding MMS relay/server.

5.9.1.5   Prepaid subscribers

The primary strategy that mobile operators have used to increase
subscriber numbers is to offer phones on a prepaid plan. Users pay
money over their phone into a network-based account, either
directly or by buying vouchers, and some of the value credited is
used up each time a phone call or other chargeable transaction is
made. No monthly subscription fees or other recurring charges are
made. At first, prepaid credit expired after a certain period, but
most operators have now abandoned the expiry period. As part
of their efforts to boost ARPU, operators are now trying to

reduce the percentage of their subscriber base on prepaid plans, although prepaid levels are still very high (e.g., between 60 and 80% of European mobile users are on a prepaid plan). So, in order to encourage widespread usage, operators must make MMS available to prepaid users (Turnbull and Bond, 2002). A number of vendors have announced prepaid support in their MMSC products, including LogicaCMG, Comverse and Ericsson.

5.9.1.5.1   Support for prepaid service in MMS

An MMS relay/server could support the prepaid concept: a prepaid customer may be charged for submitting or retrieving MMs/abstract messages. In the submission case the originator MMS relay/server may first ascertain that the originator of the MM/abstract message is a prepaid customer. The MMS relay/server may then initiate a credit check, while further processing of the MM/abstract message is put on hold. If the customer's credit is insufficient to submit this particular MM/abstract message, the originator MMS relay/server could reject it, after checking several criteria:

- size of the MM;

- content type;

- settings of information elements;

- type of abstract message.

In case an MM/abstract message cannot be accepted, the originator MMS relay/server will respond with an appropriate status value to the submit request. The MMS UA should bring this information to the user's attention. Of course, if an MM/abstract message is accepted it is further processed by the MMS relay/server.

In the retrieving case the recipient MMS relay/server may first ascertain that the recipient of the MM/abstract message is a prepaid customer. The MMS relay/server may then initiate a credit check for the particular customer, which may be performed at the time the MM/abstract message arrives at the recipient MMS relay/server. Based on the result, the MMS relay/server may reject or accept the MM/abstract message. If it is accepted (with or without checking) the MMS relay/server may perform a credit check at the time the MMS UA sends a retrieve request (the same criteria as in the sending case may be checked). In case an MM/ abstract message cannot be retrieved because the customer's account balance is too low, the recipient MMS relay/server may respond with an appropriate status value to the retrieve request. The MMS UA should bring this information to the user's attention.

For this to be successful, there would need to be close integration between the billing system and the MMS relay/MM Server. An Open Services Access (OSA) billing interface could be used between the MMS relay/MMS server to interrogate the billing system and carry out the credit check. However, this is still difficult to integrate.

5.10 OSA CHARGING, ACCOUNTING INTERFACES USED IN MM8

One of the most important, yet usually undefined, areas in tele-communications systems is billing, charging and accounting. A system that does not have adequate interfaces into billing systems will be rendered useless. The flexibility of rating and billing systems will also determine an operator's competitiveness in a market where rating innovation is important. Operators' time to market with a new idea is determined by the adaptability of the systems in place. In mobile telephony, of which MMS is a subset, the concept of accounting, sharing revenues between operators and service providers is also very important.

The 3G standards define a concept called Open Service Access (OSA: 3GPP, 2002e), which specifies a number of interfaces to allow systems to interface with each other in a well-defined manner. One set of interfaces that are relevant to MMS concern billing, charging and accounting (3GPP, 2002f, g).

5.10.1 Introduction to Open Service Access

In order to be able to implement future applications/end-user services that are not yet known today, a highly flexible framework for services is required. The OSA enables applications to implement services that make use of network functionality, which is offered to applications in terms of a set of Service Capability Features (SCFs). These SCFs provide the functionality of network capabilities that are made accessible to applications through the standardized OSA interface, on which service developers can rely when designing new services (or enhancements/variants of those already existing).

The aim of OSA is to provide a standardized, extendible and scalable interface that allows the inclusion of new functionality in the network in future releases with minimum impact on the applications using the OSA interface. The network functionality offered to applications is defined as a set of SCFs in the OSA API (Application Programming Interface), which are supported by different Service Capability Servers (SCSs). These SCFs provide access to network capabilities on which application developers can rely when designing new applications (or enhancements/variants of those already existing). The different features of different SCSs can be combined as appropriate. The exact addressing (parameters, type and error values) of these features is described in terms of stage 3 descriptions, which (defined using the OMG [Object Management Group] Interface Description Language) are open and accessible to application developers who can design services in any programming language, while the underlying core network functions use their specific protocols.

The standardized OSA API is secure: it is independent of vendor-specific solutions and programming languages or operating systems, etc. used in the service capabilities. Furthermore, the OSA API is independent of the location within the home environment where service capabilities are implemented and independent of supported service capabilities in the network.

For application developers to rapidly design new and innovative applications, an architecture with open interfaces is imperative. By using object-oriented techniques (e.g., CORBA, SOAP [Simple Object Access Protocol], etc.), it is possible to use different operating systems and programming languages in application servers and service capability servers. The service capability servers act as gateways between the network entities and the applications. The OSA API is based on lower layers using mainstream information technology and protocols. The middleware and protocols (e.g., CORBA/IIOP [Internet Inter-ORB Protocol], SOAP/XML, other XML-based protocols, etc.) and lower layer protocols (e.g., TCP, IP [Internet Protocol], etc.) should provide security mechanisms to encrypt data (e.g., TLS [Transport Layer Security], IPSEC [IP SECurity], etc.).

5.10.2 Overview of Open Service Access

OSA consists of three parts (see Figure 5.7):

- *Applications* (e.g., VPN [Virtual Private Network], conferencing, location-based applications) These applications are implemented in one or more application server.

- *Framework* This provides applications with basic mechanisms that enable them to make use of the service capabilities in the network. Examples of framework functions are authentication and discovery. Before an application can use the network functionality made available through SCFs, authentication between the application and framework is needed. After authentication, the discovery function enables the application to find out which network SCFs are provided by the SCSs. Network SCFs are accessed by the methods defined in OSA interfaces;

- *Service Capability Servers* These provide the applications with SCFs, which are abstractions from underlying network functionality. Examples of SCFs offered by SCSs are Call Control and User Location. Similar SCFs capability features may possibly be provided by more than one SCS (e.g., Call Control functionality

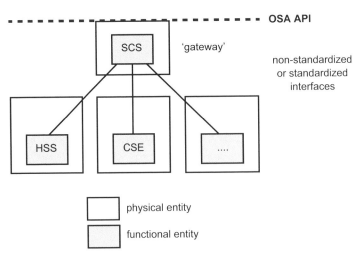

Figure 5.7 Overview of Open Service Access

API = Application Programming Interface; SCS = Service Capability Server; HSS = Home Subscriber Server; CSE = Circuit Switched Environment

might be provided by SCSs on top of CAMEL (CAMEL = Customized Application for Mobile network Enhanced Logic) and MExE [Mobile Station Application Execution Environment]).

OSA SCFs are specified in terms of a number of interfaces and their methods. The interfaces are divided into two groups:

- framework interfaces;
- network interfaces.

Note that the CAMEL Service Environment does not provide the service logic execution environment for applications using the OSA API, since these applications are executed in application servers.

SCSs that provide the OSA interfaces are functional entities that can be distributed across one or more physical nodes: for example, the User Location interfaces and Call Control interfaces might be implemented on a single physical entity or distributed across different physical entities. Furthermore, an SCS can be implemented on the same physical node as a network functional entity or in a separate physical node: for example, Call Control interfaces might be implemented on the same physical entity as the CAMEL protocol stack (i.e., in the CSE) or on a different physical entity. Several options exist:

- *Option 1* OSA interfaces are implemented in one or more physical entity, but remain separate from physical network entities. Figure 5.8 shows the case where OSA interfaces are

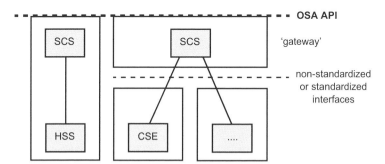

Figure 5.8 SCSs and network functional entities implemented in separate physical entities

SCS = Service Capability Server; OSA = Open Service Access; API = Application Programming Interface; HSS = Home Subscriber Server; CSE = Circuit Switched Environment

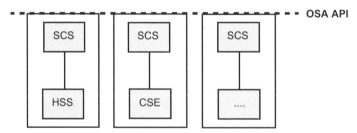

Figure 5.9 SCSs and network functional entities implemented in separate physical entities (SCSs distributed across several ``gateways'')

OSA = Open Service Access; SCS = Service Capability Server; HSS = Home Subscriber Server; CSE = Circuit Switched Environment; API = Application Programming Interface

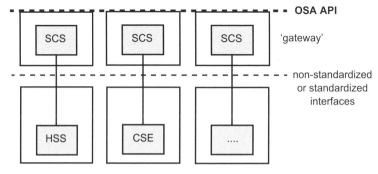

Figure 5.10 SCSs and network functional entities implemented in the same physical entities

OSA = Open Service Access; SCS = Service Capability Server; HSS = Home Subscriber Server; CSE = Circuit Switched Environment; API = Application Programming Interface

implemented in one physical entity (called "gateway" in the figure) and Figure 5.9 shows the case where the SCSs are distributed across several "gateways".

- *Option 2* OSA interfaces are implemented in the same physical entities as the traditional network entities (e.g., HSS, CSE) (see Figure 5.10).

- *Option 3* Option 3 is a combination of options 1 and 2 (i.e., a hybrid solution).

Note that in all cases there is only one framework, which may reside within one of the physical entities containing an SCS or in a separate physical entity. From the application point of view, it will make no difference which implementation option is chosen (i.e., in all cases the same network functionality is perceived by

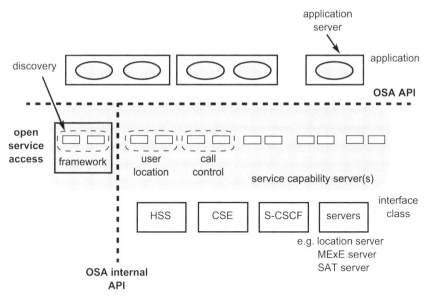

Figure 5.11 Hybrid implementation (combination of options 1 and 2)
OSA = Open Service Architecture; API = Application Programming Interface; HSS = Home Subscriber Server; CSE = Circuit Switched Environment; CSCF = Call Server Control Function; MExE = Mobile Stations Applcation Execution Environment; SAT = SIM Application Toolkit

the application). Applications will always be provided with the same set of interfaces and have common access to framework and SCF interfaces. It is the framework that will provide the applications with an overview of available SCFs and the knowledge of how to make use of them.

5.10.3 Basic mechanisms in the Open Service Access

This subsection explains which basic mechanisms are executed in OSA prior to offering and activating applications. Some mechanisms are applied only once (e.g., establishment of service agreement) while others are applied each time a user subscription is made to an application (e.g., enabling the call attempt event for a new user). The basic mechanisms taking place between the application and the framework are:

- *Authentication* Once an offline service agreement exists, the application can access the authentication function. The authentication model of OSA is a peer-to-peer model. The application must authenticate the framework and vice versa. The application must be authenticated before it is allowed to use any other OSA function.

- *Authorization* Authorization is distinguished from authentication in that the former is the action of determining what a previously authenticated application is allowed to do. Authentication must precede authorization. Once authenticated, an application is authorized to access certain SCFs.

- *Discovery of framework functions and network SCFs* After successful authentication, applications can obtain available framework functions and use the "Discovery" function to obtain information on authorized network SCFs. The "Discovery" function can be used at any time after successful authentication.

- *Establishment of service agreement* Before any application can interact with a network SCF, a service agreement must be established, which may consist of an offline (e.g., by physically exchanging documents) and an online part. The application has to sign the online part of the service agreement before it is allowed to access any network SCF.

- *Access to network SCFs* The framework must provide access control functions to authorize access to SCFs or service data for any API method from an application, with the specified security level, context, domain, etc.

The basic mechanism taking place between the framework and the SCF is:

- *Registration of network SCFs* SCFs offered by an SCS can be registered at the framework. In this way the framework can inform applications upon request about available SCFs (Discovery) (e.g., this mechanism is applied when installing or upgrading an SCS).

The basic mechanism taking place between the application server and the SCS is:

- *Request of event notifications* This mechanism is applied when a user has subscribed to an application and that application needs to be invoked on receipt of events from the network related to the user. For example, when a user subscribes to an incoming call-screening application, the application needs to be invoked when the user receives a call. So for this scenario the application would be notified when a call set-up is performed.

5.10.4 Handling of end-user-related security

Once OSA's basic mechanisms have ensured that an application has been authenticated and authorized to use network SCFs, it is also important to handle end-user-related security aspects which consist of:

- *End-user authorization to applications* This limits end-user access to just those applications they have subscribed to.

- *Application authorization to end-users* This limits usage of network capabilities by applications to authorized (i.e., subscribed) end-users.

- *End-user privacy* This allows the user to set privacy options.

5.10.4.1 End-user authorization to applications

An end-user is authorized to use an application only when he or she has subscribed to it. In the case where the end-user has subscribed to the application before the application accesses the network SCFs, then the subscription is part of the Service Level Agreement (SLA) signed between the Home Environment (HE) and the HE-VASP (Value Added Service Provider). After the application has been granted access to network SCFs, subscriptions are controlled by the HE. Depending on the identity of an authenticated and authorized end-user, the HE may use any relevant policy to define and possibly restrict the list of services to which a particular end-user can subscribe. At any time, the HE may decide, unilaterally or after agreement with the HE-VASP, to cancel a particular subscription. Service subscription and activation information need to be shared between the HE and the HE-VASP, so that the HE-VASP knows which end-users are entitled to use its services. Appropriate online and/or offline synchronization mechanisms (e.g., SLA renegotiation) can be used between the HE and the HE-VASP (not specified in OSA release 5). End-to-end interaction between a subscribed end-user and an application may require the usage of appropriate authentication and authorization mechanisms between the two that are independent of the OSA API, and therefore outside the scope of OSA standardization.

5.10.4.2 Application authorization to end-users

The HE is entitled to provide service capabilities to an application with regard to a specific end-user if the following conditions are met:

1. the end-user has subscribed to the application;

2. the end-user has activated the application;

3. the usage of this network service capability does not violate the end-user's privacy settings (see next subsection).

The SCS ensures that the above conditions are met whenever an application attempts to use an SCF for a given end-user and to respond to the application accordingly (possibly using relevant error parameters).

5.10.4.3 End-user privacy

The HE may permit an end-user to set privacy options. For instance, it may permit the end-user to decide whether his or her location may be provided to third parties, or whether he or she accepts information to be pushed to his or her terminal. Such privacy settings may have an impact on the ability of the network to provide SCFs to applications (e.g., user location, user interaction). Thus, even if an application is authorized to use an SCF, the end-user has subscribed to this application and this application is activated, privacy settings may still prevent the HE from fulfilling an application request. The SCS ensures that a given application request does not violate an end-user's privacy settings or that the application has relevant privileges to override them (e.g., for emergency reasons).

5.10.5 Account management SCF

Each aspect of the account management SCF follows this sequence:

- sequence diagrams give the reader a practical idea of how each SCF is implemented;

- the class relationships clause shows how each of the interfaces that are applicable to the SCF relate to one another;

- the interface specification clause describes in detail each of the interfaces shown within the class diagram part;

- State Transition Diagrams (STDs) show the progression of internal processes either in the application or the gateway;

- the data definitions section shows in detail each of the data types associated with the methods within the classes.

5.10.6 Charging SCF

This subsection describes each aspect of the charging SCF. The order is as follows:

- sequence diagrams give the reader a practical idea of how each SCF is implemented;

- the class relationships clause shows how each of the interfaces that are applicable to the SCF relate to one another;

- the interface specification clause describes in detail each of the interfaces shown within the class diagram part;

- STDs show transitions between states in the SCF, both of which are well defined: either methods specified in the interface specification or events occurring in the underlying networks cause state transitions;

- the data definitions section shows each of the data types associated with the methods within the classes in detail.

5.10.7   Sequence diagrams

5.10.7.1   Reservation/Payment in parts

The sequence diagram (Figure 5.12) illustrates how to request a
reservation and how to charge a user from the reserved amount
(e.g., to charge a user for a streamed video that lasts 10 minutes and
costs a total of $2.00). Operations and interfaces that do not provide
rating are employed throughout this sequence diagram. Let us
assume the application has already discovered the charging SCF.
As a result, the application received an object reference pointing to
an object that implements the ChargingManager interface. Operations that handle units are used in exactly the same way, except that
the amount of application usage is indicated instead of a price:

1. The application creates a local object implementing the App-
   ChargingSession interface. This object will receive response
   messages from the ChargingSession object.

2. The application opens a charging session and a reference to a
   new or existing object-implementing ChargingSession is
   returned together with a unique session ID.

3. In this case a new object is used.

4. The application requests the reservation of $2.00.

5. Assuming the criteria for requesting a reservation are met (the
   application provider has permission to charge the requested
   amount, the charged user has agreed to pay the requested
   amount), the amount is reserved in the session. At this point
   the application provider knows that the network operator will
   accept later debit requests up to the reserved amount. So, the
   application may start serving the user (e.g., by sending the
   MMS notification message).

6. The successful reservation is reported back to the application.

After half of the video has been sent to the user, the application
may choose to capture half of the price already:

7. The application requests permission to debit $1.00 from the
   reservation.

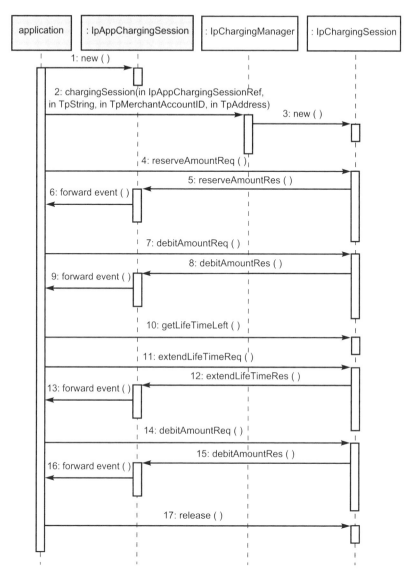

Figure 5.12 Message sequence diagram (1)
Source: 3GPP (2002f)

8. The successful debit is reported back to the application.

9. The acknowledgement is forwarded to the application.

10. The application checks whether the remaining lifetime of the reservation will cover the remaining 5 minutes of video – let us assume it does not.

11. The application asks the ChargingSession object to extend the lifetime of the reservation.

12. Assuming that the application provider is allowed to keep reservations open for longer than 10 minutes, the extend Life-TimeReq function will be honoured and confirmed properly.

13. The confirmation is forwarded to the application.

14. When the complete video has been transmitted to the user without errors, the application charges another $1.00.

15. The ChargingSession object acknowledges the successful debit at the IpAppChargingSession callback object.

16. The AppChargingSession object forwards the acknowledgement to the application.

17. Since the service is complete, the application frees all resources associated with the reservation and session.

5.10.7.2 Immediate charge

The sequence diagram (Figure 5.13) illustrates how immediate charging is used. Assume a WAP gateway that charges the user $0.01 per requested Uniform Resource Locator (URL). Since it is acceptable to lose one tick worth $0.01, no prior reservations are made. The WAP gateway sends an immediate debit for each re-quested URL, and should a payment have a result failure the user is disconnected. Operations that handle units are used in exactly the same way, except that the amount of application usage is indicated instead of a price:

1. The application creates a local object implementing the AppChargingSession interface. This object will receive response messages from the ChargingSession object.

2. The application orders the creation of a session. No new object is created for charging session handling in this sample implementation.

3. The application requests permission to charge the user $0.01.

4. The payment is acknowledged.

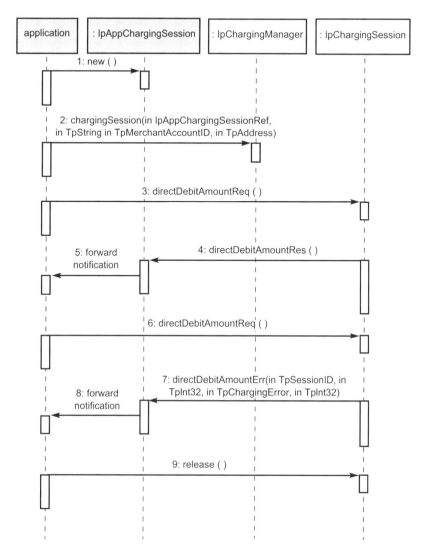

Figure 5.13 Message sequence diagram (2)
Source: 3GPP (2002f)

5. The acknowledgement is forwarded to the application.

6. The application requests permission to charge the user $0.01.

7. The payment is reported to fail.

8. The failure report is forwarded to the application.

Note that Steps 3–5 and 6–8 may be repeated for as long as you want and in any order you want.

9. The application releases the session.

5.10.8 Inband MMS detection for charging

One conceptual method that could be used to charge for MMS usage is the idea of an inband device that intercepts packets and charges accordingly. This method would avoid interfaces with the MMSC having to be developed, thus reducing the complexities involved. One such device is the ProQuent Systems Mobile Services Switching Point (MSSP) 2800 product family, a carrier-class programmable IP service switch for the wireless data market (ProQuent Systems, 2003). The MSSP models the Service Switching Point (SSP) deployed in Intelligent Network (IN) voice networks, but is designed for mobile IP data applications. The MSSP separates the control and data transport functions to give operators full control over a user's data session. As the MSSP inspects the context and state of IP flows, it involves service logic from applications by using a patented IP Detection Point^{TM} technology. This functionality allows operators to develop a wide range of services and service capabilities. The MSSP enhances the functionality of existing network elements by providing the following features:

• *Enhanced charging information* This provides detailed information for each user session and provided service, so billing can be performed intelligently based on content and service provided, not just by byte count.

• *Open application interfaces* These expose control points in user sessions that can be accessed through applications. These control points enable operators to extend control over subscriber flows in order to manage chargeable content and to provide a better user experience.

• *Performance monitoring* This captures real-time and historical performance information on user data sessions, including usage by application and service.

- *Centralized management location* This connects with management, billing, performance, provisioning and application servers from a central location between the wireless infrastructure and the Internet.

MSSP uses the content-based charging model, which offers operators the flexibility to charge based on a variety of parameters, such as content, destination, time and bytes transferred. This model can provide separate charges for transport (number of bytes) and content during a data session. By using the MSSP to separate transport and content, traditional voice-based charging models can be reproduced for the wireless data market and new charging models can be introduced:

- *Calling party pays* Content-based charging allows operators to charge the calling party for the bytes transferred and the content accessed (equivalent to an MMS mobile origination service).

- *Called party pays* Content-based charging allows operators to offer a reverse charging application where the called party (such as a content provider) is charged for the bytes transferred and the content accessed (equivalent to a MMS mobile termination service).

- *Third party pays* More sophisticated charging models can be implemented where a third party (someone other than the called party) is charged for the bytes transferred and the content accessed.

- *Multiple parties pay* The MSSP opens up a new category of charging that involves multiple paying parties. A single subscriber session may incur charges for the calling party, called party and third party (e.g., an advertiser).

5.10.9 Fraud

Certain elements of society will always try to obtain services for free or even to obtain funds through illegal means using the services. MMS services are susceptible to these kinds of potential revenue loss. In order to prevent revenue losses, operators and service providers will need to have fraud or revenue assurance systems in place. It is difficult to predict where any losses will occur, but it is important to have the systems in place to detect them when it happens. One of the methods used to counter the threat is to have systems employing artificial intelligence techniques to monitor charging records for fraudulent attacks on the system. Such a system is provided by a company called Azure (www.azure.co.uk) that specializes in revenue assurance.

5.10.10 Sample charging scenario

MMSC calculates the cost of the MMS and asks a user to confirm it. The cost includes storage and, because the photo is high resolution and exceeds the standard threshold of 100KB, an excess charge is requested. The user confirms and the photo is sent. Later, an item on her bill will read "Hi-Res Photocard, e9", with other datestamp and destination information (Machin, 2002). Let us take the example of an internetwork photocard and look at the steps involved:

1. This is the content creation step. John creates a photocard using a camera-enabled mobile T device, which allows him to take a picture and add a text message. Using inbuilt options on the device he opts to send it to a friend, Jane, on another network. The operator offers him some choices: the operator can simply send the picture, or he can send it and store it on his network clipboard, rather than occupying valuable storage space on his device. Another option would be to simply store it for later access from a networked PC. John opts to send and store.

2. The MMS has arrived at the destination network's MMSC. The receiver is notified that there is a photocard waiting for her and invited to download. She can accept this invitation, leave it in storage for later access via PC or forward it to another device or mailbox. All of these options would be charged differently. Jane opts to download to her device – she wants to see it straight away – but also to store it on her local clipboard (later she wants to access it from her PC at home). The MMSC calculates a charge for storage based on the size of the message and informs her. Jane's transaction with the costed message is delivered to her free of charge and the only charge is the small payment for clipboard storage (note that the operator could charge for delivery if

this was thought acceptable to the customer – but not in a "calling party" model).

3. This has been an internetwork MMS. Jane's network will want to recoup from John's network the cost of delivering the MMS. The two networks will have agreed an interconnect tariff for the exchange of MMSs. This could be based on a number of criteria including date and time, MMS type and so on, but a simple tariff based on MMS size has been assumed. Both networks will maintain a record of the exchange and will settle later through normal intercarrier settlement processes.

5.10.11   MMS billing and accounting

Let us now look at some of the key issues in billing for MMSs and their impact on the billing environment. MMS data collection CDRs will be generated at the MMS relay and at the MMS server. Note that these require a mediation capability: an interaction between the MMS relay and server to collect CDRs. CDRs must contain all available charging information (mediation may be combined as a single entity). Data that can be used for charging purposes would include:

- MMS type;
- MMS size;
- time and date stamp;
- MMS send/receive indicator;
- sender identification;
- receiver identification.

5.10.12   Advice of charge

An important feature of the MMS standard is advice of charge (3GPP, 2002j). This is a service that notifies the MMS recipient of the cost of receiving it. This is important when a subscriber pays to upload content. Although, it may be desirable not to inform subscribers of the cost until the bill arrives, they are more likely to use a service if the cost is presented to them prior to accepting it. The charge is based on the subscriber's selection and the MMS receiver may opt to:

- accept delivery of the MMS;
- decline/defer delivery of the MMS.

Flexible rating based on subscriber action:

- delete the MMS without viewing
- forward the MMS to an alternative viewing terminal – a PC or Personal Digital Assistant (PDA)

5.10.13 Changing models

There are a number of changing models available:

- Daily news may be provided at €30 per month, with no individual charging of MMS.
- Instant messages may be free, up to 10KB volume/message.
- Multiple CDR charging (e.g., charging for a game session).
- Single charging event covering multiple messages.
- Where the operator does have visibility and responsibility for MMS content (e.g., P2P or inbound MMS).
- Rating based on network resource usage (message size or volume).
- Interconnect charging – a need to establish interconnect for inbound MMSs and an interconnect system MMS tariff. This would probably be charged based on accumulated MMS volumes.

5.10.14 Support for reply-charging in MMS

The MMS UA, relay/server and VASP that are connected to an MMS relay/server can support supply charging. The user of the MMS (the originator MMS UA or VASP) may be able to take control of charging for the sending of a reply-MM (to their submitted MM) from the recipient(s). Therefore, the originator of an MM (either the MMS UA or the VASP) should be able to mark the MM as reply-charged. The originator's MMS relay/server has the option to accept the user or VASP settings for reply-charging and should be able to convey feedback to the originator. It should be possible to take control of charging for reply-MMs from different recipients.

The recipient should be notified if she is not charged for a reply-MM to this particular MM. However, it should be made clear that reply-charging does not apply to a reply-MM to an original MM (i.e., this is free of charge) and that retrieval of the original MM marked as reply-charged is not free of charge. Both the originator and the recipient MMS relay/server will be able to ensure that no more than one reply-MM per recipient is charged to the originator. The MMS UA should let the user know if an MM has already been replied to. The request for reply-charging will not be passed on to the recipient:

- if the recipient is not known to belong to an MMSE peer entity, or
- in the case the MM is forwarded.

The following behaviour is expected to support reply-charging in MMS. As part of the submission of an MM, the MM originator (either the MMS UA or the VASP) may indicate a willingness to pay the charge for one reply-MM per MM recipient. In this case the originator MMS UA or originator VASP:

- will indicate the sender's willingness to pay the charge for one reply-MM per MM recipient;
- may define a reply-charging limitation request (e.g., may specify both the latest time of submission or the maximum size for the reply-MMs).

In response to the MM submission, the originator MMS relay/ server will inform the MM originator (either the MMS UA or the VASP) whether or not it accepts:

- the originator's request for reply-charging in the original MM;
- the reply-charging limitations set by the originator (either the MMS UA or the VASP) in the original MM.

On receipt of an MM from an originator (either the MMS UA or the VASP), the originator MMS relay/server:

- may provide reply-charging limitations (i.e., it may further limit the MMS UA or VASP's settings for reply-charging);

- will let it be known whether or not a reply-MM is requested in an unaltered state when routing the original MM toward the MM recipient(s), if the peer entity is known to be the same MMS relay/server;

- will pass the reply-charging limitations for the reply-MM when routing the original MM toward the MM recipient(s), if the peer entity is known to be the same MMS relay/server.

If the MM recipient has requested the original MM to be forwarded to some other address, the recipient MMS relay/server will not pass any information about the reply-charging request on to the addressee(s) of the forwarding request. If reply-charging has been requested by the MM originator (either the MMS UA or the VASP), the recipient MMS relay/server:

- should inform the recipient MMS UA by means of the MM notification and on MM delivery that the MM originator is willing to pay for a reply-MM to this original MM;

- may notify the recipient about the reply-charging limitations set by the originator (e.g., the latest time of submission of a reply-MM to the original MM).

When a user intends to send a reply-MM to the MM originator (either the originator MMS UA or the VASP), the recipient MMS UA (i.e., the originator MMS UA of the reply-MM):

- will mark the MM as a reply-MM;

- will provide the message ID of the original MM that it is replying to (if it is the reply-MM);

- will submit the reply-MM to the recipient MMS relay/server;

- may be able to indicate to the user whether this MM has already been replied to;

- may be able to indicate to the user whether the reply-charging limitations cannot be met.

On submission, the recipient MMS relay/server:

- will reject the reply-MM submission attempt and should convey this information back to the recipient MMS UA (the originator MMS UA of the reply-MM), if the reply-MM submission attempt does not meet the limitations set by the originator (either the MMS UA or the VASP);

- will be able to uniquely map the reply-MM to the original MM.

5.10.14.1 Use-case for reply-charging

The following detailed, sample use-case of reply-charging describes the case when MMS UA A and MMS UA B belong to the same MMSE. MMS UA A is the sender of the reply-charged MM and MMS UA B is the recipient of the reply-charged MM. In the message flow when reply-charging, user A produces an MM and marks it "reply-charged" before it is submitted to the MMS relay/server. The MMS Relay/Server notes that user A is willing to pay for a reply-MM to this particular MM and notes the message identification of the original MM and the originator's limitations. The MM is retrieved by user B in accordance with user B's user profile. This might imply charges for user B when retrieving the MM. User B retrieves the original MM and discovers that the first reply to this message (that is accepted by the service provider) will be paid by user A. User B creates an answer and the MMS UA B marks it as a reply-MM and submits it to the MMS relay/server. The MMS relay/server identifies this MM as a reply to the original MM and checks the originator's limitations. If the MMS relay/server accepts the reply, the reference set before (as described in transaction 1 of Figure 5.14) is deleted and user A is billed for transaction 3. User A retrieves the reply-MM and eventually is billed for transaction 4. The case of reply-charging where the originator MMS UA is actually the MMS VAS application (using the MM7 reference point)

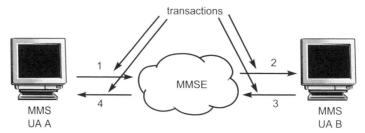

Figure 5.14 Message flow in the event that reply-charging needs to be applied

behaves in the same way as the use-case of two MMS UAs in the same MMSE.

5.11 INTERNETWORK INTEROPERABILITY USING THE MM4 INTERFACE

Mobile text messaging has largely been dominated by mobile operators, with SMS phones sending messages to other SMS phones. MMS, with its foundation on open, Internet-based standards, will demand a greater amount of interoperability with other types of device (such as PDAs or PCs) and other messaging services (such as IM or Internet email). There are two reasons for this:

• In the early stages of a developing MMS market, when few people have MMS phones, those who do have them will want to be able to send messages to people who do not. This means that it must be possible to deliver at least some of the message content to existing devices (such as PC-based web browsers, email accounts, WAP phones and SMS phones).

• As the MMS market matures, other messaging services – principally wireless email and wireless instant messaging – will have evolved sufficiently that an increasing number of users will be able to access them via wireless devices at least some of the time. As SMS has shown us, one of the most important factors of a messaging service's value is the ability to communicate within a wide community, regardless of network operator. By encompassing email and instant messaging users, service providers will greatly expand the community, accessible to their MMS users.

Therefore, another important difference between SMS and MMS business models is that more gateways will be needed in the operator's network: interconnect relationships will be needed with providers of messaging services who are not mobile operators.

The current 3GPP specification does not directly tackle the issue of VASP interconnect with multiple operator networks. Currently, a VASP must have an agreement with each mobile operator, to establish a VASP identifier and ensure the billing mechanisms are in place. This will make it difficult for content providers and VASPs to deliver content on a global basis. There are also some un-answered questions, principal among which is: Will operators and service providers permit incoming Internet email to mobile phones? This scenario has been permitted on a limited scale with email to SMS, often with advertising added to the message or a limit on the number of emails that can be sent in a day. The O_2 mmail service allows email to SMS and charges the receiver, not the sender; so, it is possible that some of the problems can be resolved.

5.11.1 The MM4 interface

The MMSC architecture (Figure 5.15) specifies an interface called MM4 to connect foreign MMSCs and can be implemented using SMTP, to transport Internet email with a Multipurpose Internet Email Extension (MIME) attachment, which contains the MM and an MMS header (for event processing and interconnect charging). According to the MMSC architecture the last MMSC in a chain to receive an MMS will send the notification (an MMS notif.ind, which

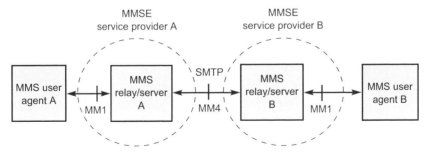

Figure 5.15 MMSE interoperability between network elements

MMSE = Multimedia Messaging Service Environment; SMTP = Simple Mail Transfer Protocol. Source: 3GPP (2002a)

is a WAP-PUSH message formatted in a binary SMS) to the recipient, chargeable to this MMSC operator (at a typical cost of 1p). In a typical P2P MMS, the sender will pay the cost to transmit the message; so, there must be a way for the receiving MMSC to charge the sender's MMSC operator who in turn must be able to charge the sender. The billing issue is solved by collecting CDRs in each MMSC and reconciling them between operators.

The user's phone has an MMSC setting (which is effectively a URL to a Web server) that is really only relevant in current specifications when you send an MMS message. The phone always connects to the configured MMSC to send a message. When receiving an MMS, essentially the MMS notification to a mobile device is sent as a special SMS message (often two concatenated, binary SMS messages). The sending and receiving of the MMS content takes place over the mobile network (usually over IP or in Europe over GPRS [General Packet Radio Service], but it is possible to use CDMA [Code Division Multiple Access], PHS [Personal Handyphone System] or other network technologies). So, once an MM has been received, the phone receives the SMS notification. This SMS (WAP-PUSH message) contains a pointer to the URL that is hosting the MMS content. Whether or not the phone can receive that message depends on whether or not the active profile for MMS points to a configuration set that allows the phone to retrieve the URL, as the content is fetched through a WAP gateway. What comes into play here is that the active GPRS Access Point Node (APN) setting can limit which WAP gateway the phone connects to, and the WAP gateway setting can limit which content servers it connects to (in the case of MMS the content server is usually the MMSC). If connecting through an operator WAP gateway, it should be possible to fetch content from both. If you're going through an independent WAP gateway, it depends on whether or not operators have firewalls around their MMSCs that block message retrieval from outside their network.

5.11.2 Resolving the recipient's MMSE IP address

For those recipients that appear in an MM and belong to an external MMSE, the originator MMS relay/server has to send the message to each of their MMS relay/servers using SMTP. The MMS relay/ server has to resolve the recipient's MMS relay/server domain name to an IP address (e.g., using the Domain Name System

[DNS]), based on the recipient's address. Mapping the recipient's address, as in the case of MS-ISDN [MS Integrated Services Digital Network] (E.164) addressing, to the recipient's MMS relay/server (if the MM recipient belongs to another MMSE can be done) using the DNS-ENUM (Electronic Numbering) protocol.

An ENUM-based solution is not the only answer; however, it is expected that MMS service providers or network operators may use solutions for their particular needs, which may include **static tables** or other look-up methods. An alternative to such a look-up method, based on MS-ISDN to IMSI (International Mobile Subscriber Identity) look-up, is described in the section below.

5.11.3 Reformatting sender and recipient addresses to Full Qualified Domain Name (FQDN) format

When delivering a message from an MMSE to another MMSE, both the sender and the recipient addresses will be extended to include the FQDN to enable transport over SMTP. This FQDN format will be used in the MM4 reference point. Although it is required that FQDN format address is used in "MAIL FROM:" and "RCPT TO:" commands in SMTP, it is not necessary that originator and recipient addresses in the "From:" or "To:" fields are reformatted in FQDN format.

5.11.3.1 DNS-ENUM recipient MS-ISDN address resolution

For those recipient MS-ISDN addresses that appear in an MM and belong to an external MMSE, the originator MMS relay/server will resolve them to a routable RFC (Request For Comments) 2822 address that will be used in the "RCPT TO:" SMTP subsequent commands.

5.11.3.1.1 DNS-ENUM recipient MS-ISDN address resolution procedure

The originator MMS relay/server will ensure that the recipient address (MS-ISDN) complies with the E.164 address format and includes the "+" character. In the case of national or local addressing scheme (e.g., just an operator code followed by a number), the MMS relay/server will convert the national or local number to an E.164 address format:

+30-697-123-4567

The originator MMS relay/server will remove all characters with the exception of digits and put dots (".") between each digit and reverse the order. The resulting subdomain will be converted to an FQDN by appending an appropriate string, which depends on the administrative control of the ENUM implementation:

    7.6.5.4.3.2.1.7.9.6.0.3.e164.arpa   (public top-level domain),
    7.6.5.4.3.2.1.7.9.6.0.3.e164.gsm   (private top-level domain),
    7.6.5.4.3.2.1.7.9.6.0.3.e164.gprs   (private top-level domain), etc.

The resulting FQDN together with the string (E.164 number) in the form as specified above will be used as the input to the NAPTR algorithm to find an applicable Resource Record (RR) (see the next subsection) by the originator MMS relay/server. The output may result in one of the following cases:

• E.164 number not in the numbering plan – the originating MMS relay/server will invoke an appropriate address resolution exception handling procedure (e.g., send a message to the originating MMS UA reporting the error condition).

• E.164 number in the numbering plan, but no URIs (Uniform Resource Identifiers) exist for that number. The originating MMS relay/server will invoke an appropriate address resolution exception handling procedure (e.g., send a message to the originating MMS UA reporting the error condition, perform the necessary conversion and route the message forward to the recipient via MM3, etc.).

• E.164 number in the numbering plan, but no MMS URIs exist for that number (MMS URIs are of the form "mms:mailbox" and they are defined in the MMS RR subsection). The originating MMS relay/server will invoke an appropriate address resolution exception handling procedure (e.g., send a message to the originating MMS UA reporting the error condition, perform the necessary conversion and route the message forward to the recipient via MM3 using the appropriate URI based on the Service field, etc.).

- DNS ENUM service unavailable. The originating MMS relay/ server will invoke an appropriate address resolution exception handling procedure (e.g., send a message to the originating MMS UA reporting the error condition, store the message in the queue and retry at a later time, etc.).

5.11.3.1.2 MMS Resource Record (RR)

The key fields in the NAPTR RR are the Domain, TTL (Time To Live), Class, Type, Order, Preference, Flags, Service, Regexp and Replacement. In particular, for this release the following fields are further specified as follows:

```
Service = "mms+E2U"
Regexp = "!^.*$!mms:mailbox!"
```

Remember: the MMS URI is of the form "mms:mailbox". The following is an example of the NAPTR RRs that are associated with the FQDN derived from the recipient MS-ISDN address (+306971234567):

```
IN NAPTR 100 10 "u" "sip+E2U" "!^.*$!sip:Mary.Smith@sip.cosmote.gr!"
IN NAPTR 100 11 "u" "mms+E2U"

"!^.*$!mms:+306971234567/TYPE=PLMN@mms.cosmote.gr!" .
            IN NAPTR 101 10 "u" "mailto+E2U"

"!^.*$!mailto:Mary.Smith@mycosmos.gr!" .
            IN NAPTR 102 10 "u" "mailto+E2U"
"!^.*$!mailto:MaryS@otenet.gr!" .
            The +306971234567 is converted to the following URIs:
            sip:Mary.Smith@sip.cosmote.gr
            mms:+306971234567/TYPE=PLMN@mms.cosmote.gr
            mailto:Mary.Smith@mycosmos.gr
            mailto:MaryS@otenet.gr
```

The originator MMS relay/server will resolve the domain part of the "mailbox" of the highest precedence MMS URI to an IP address using standard DNS. For example, the highest precedence MMS URI is:

mms:+306971234567/TYPE=PLMN@mms.cosmote.gr

The domain part of the "mailbox" is mms.cosmote.gr and is

resolved (e.g., using DNS) to 10.10.0.1. The resulting IP address together with the recipient RFC 2822 address ("mailbox") will be used by the originator MMS relay/server for routing the MM forward using SMTP to the recipient MMS relay/server.

5.11.3.2   Recipient MS-ISDN address resolution based on IMSI

The alternative solution to address resolution between MMSEs is based on a MAP query, which is a signalling request to the MMSE HLR. For those recipient MS-ISDN addresses that appear in an MM and belong to an external MMSE, the originator MMS relay/server may resolve them to a routable RFC 2822 address that will be used in the "RCPT TO:" SMTP subsequent commands.

5.11.3.2.1   Recipient MS-ISDN address resolution procedure
The originator MMS relay/server determines the recipient MS-ISDN address as belonging to an external MMSE. The originator MMS relay/server will interrogate the recipient HLR for the associated IMSI by invoking standard GSM MAP operations SRI_for_SM or/and Send_IMSI. In the event of successful interrogation the originator MMS relay/server will determine the MCC (Mobile Country Code) and MNC (Mobile Network Code) and look for a matching entry in an IMSI table. The IMSI table will maintain the associations:

$$MCC + MNC \rightarrow MMSE\ FQDN$$

Subsequently, the originator MMS relay/server will be able to resolve (e.g., using standard DNS) the MMSE FQDN to an IP address for establishing the SMTP (MM4) session.

   If the recipient MS-ISDN is not known to belong to any MMSE (no entry in the IMSI table, GSM MAP error, etc.), the MMS relay/server will invoke an appropriate address resolution exception handling procedure (these procedures are not standardized). This procedure complies with the requirements for Mobile Number Portability (MNP). Figure 5.16 provides a sample message flow diagram.

   Note that, although the used GSM MAP operations are standardized operations, in some cases HLR is unable to return the correct recipient IMSI number (GSM MAP error received) due to recipient or recipient network's settings. In that case the MMS relay/server

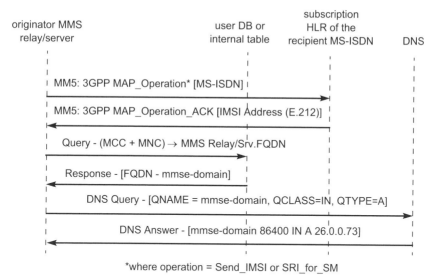

Figure 5.16 Message flow of recipient MS-ISDN address resolution based on IMSI

MMS = Multimedia Messaging Service; DB = Data Base; HLR = Home Location Register; MS-ISDN = Mobile Station Integrated Services Digital Network; Integrated Services Digitial Network; MAP = Mobile Application Part; IMSI = International Mobile Subscriber Identity; MCC = Mobile Country Code; MNC = Mobile Network Code; FQDN = Full Qualified Domain Name; DNS = Domain Name System. Source: 3GPP (2002a, p. 113)

will invoke an appropriate exception handling procedure (again these procedures are not standardized).

5.11.4 The importance of interoperability

From SMS experience, it has been learnt that the key driver of take-up is interoperator messaging capability. As long as messages can only be sent to subscribers of the same network, the appeal is limited.

Australian operator Optus Mobile claims that interoperability in SMS boosted take-up by 1,000% in April 2000 compared with April 1999 when their customers could only send messages to other Optus Mobile subscribers (see **www.vodafone.com** media centre). Other operators saw similar leaps in SMS messaging when subscribers were allowed to send messages to subscribers of other networks. It is therefore likely that the introduction of interoperator MMS messaging will provide a large boost to its take-up. However, even though the attractiveness of interoperability is undeniable, it is not easy to achieve. There are both technical and commercial issues to overcome before interoperability can become a fact of life for MMS providers.

5.11.4.1 *Technical obstacles*

The issue that is causing the most problems for MMS interconnection is physical interoperability between a pair of operator networks. Unfortunately, in their rush to adopt MMS technology, many mobile operators built and installed proprietary interfaces without waiting for the 3GPP standard to be implemented, which now makes the interconnection of MMSCs provided by different vendors difficult. LogicaCMG and other messaging infrastructure providers are working on interoperability solutions, but they have not yet arrived at a simple plug-and-play solution. In the meantime, pairs and groups of mobile operators are working on how to get their various pieces of equipment to talk to each other. The good news is that these obstacles are starting to be overcome: the weeks before December 2002 saw at least six separate announcements of MMS interconnect deals reached between operators. Further good news is that, once the issue has been overcome between a pair of operators, this can largely be extended to all the other pairs and groups of operators using the same interfaces. It is no coincidence that after the first Nokia–Ericsson pairing was introduced, several others quickly followed. All of this means that the root technical obstacles are largely being dealt with, leaving only operator-specific issues to deal with. For the moment though, technical issues are still causing some concern.

5.11.4.2 *Commercial obstacles*

In the event that agreement cannot be made promptly on commercial terms, services can still be offered without interoperation charge on an interim basis. However, if billing systems are unable to interact, the commercial viability of MMS will be affected.

Although the technical issues are being resolved, commercial issues are proving more of a challenge (this is often the case in interconnection), but at the end of the day they must and will be resolved. Where charges are not regulated – and MMS interconnection is likely to remain unregulated for some time (at least in most countries) – operators are free to negotiate for the best deal. In some cases, this will likely cause as many delays as the technical issues.

Early deployments of MMS and initial interconnect agreements have been on a gratis basis, where no payments were exchanged at all between the operators, on the assumption that traffic flows between them are fairly even. This is clearly unsustainable. The 3GPP charging architecture supports reconciliation of transferred messages between MMS environments, which will lead to per-message charging (where the operators pay each other per message that terminates on their network). Per-message charging is certainly easy to do, and most operators already operate in this way for SMS messaging. However, it has major drawbacks for MMS messages, where the size of messages might vary from one or two kilobytes to several hundred. This method would have to choose an average message size as the norm: undercompensating at some times and overcompensating at others. An alternative would be to agree price bands for messages of a certain size or content type: say, 100–300KB is band A or a single photo image is band B.

5.12 OPEN SOURCE COMPONENTS FOR BUILDING AN MMSC

Open source software has become increasingly popular in recent years. A distinction should be made between open source software and free software: open source software gives the developers and users the ability to inspect the software being used and to modify it when required, but the copyright owners of the software can still charge a fee for doing so, while free software is where it can be used ubiquitously for no fee. The Open Source Software Organization specifies rules and guidelines for the distribution, modification and use of free software.

Currently, there is no open source implementation of a complete MMSC relay/server. But, there are components available that would reduce the effort involved in constructing the system completely from scratch. In order to build the system the following capabilities would be required:

- operating system;
- Web server;
- mail transfer agent;
- WAP gateway;
- SMSC interface;
- management/OSS system;
- billing system or billing system interfaces;
- the application software that would form the actual MMS relay/ server;
- implementation of any of the other required interfaces.

Choice of operating system would seem obvious. Linux has become the open source and free operating system of choice for a large number of developers and organizations. Although still evolving, it has become the de facto standard for hosting Web servers, a testimony to its robustness when used in the high-transaction rate, 24/7 environment. There are doubts about Linux when considering it for use in a desktop environment, but this is not relevant to its use in hosting an MMSC. It has yet to achieve prime time in the telecommunications industry, for two principal reasons: first, it is not perceived to be ready for mission-critical applications that provide operators with their sole source of income (if it fails who do you sue?), and, second, the fact that a major part of a platform is free has a significant impact on the amount that a supplier can charge for systems that need to be developed. This is similar to the effects the cost of hardware have had on the cost of software. Recently, operators have adopted the approach of using general purpose hardware at a significant cost saving. Traditionally, operators demanded that all systems run on fault-tolerant hardware, which meant the development of hardware such as that used in voice switches or using specialized fault-tolerant platforms (e.g., Tandem and Stratus computers). Suppliers charged millions of dollars for voice switches and hundreds of thousands for fault-tolerant computer platforms, resulting in operators having to pay tens of thousands for computer hardware, which has an effect on the amount suppliers can charge for develped software. A similar effect is likely to occur on the cost of developed software if Linux is used.

An open source WAP gateway is provided by the Kannel Foundation. The Kannel WAP gateway is distributed under a BSD-style license, which was chosen in preference to other open source licenses because it placed the fewest limitations on what a third party can do with the source code (Wirzenius and Marjola, 2002). The Kannel WAP gateway is being developed on Linux, but can be ported to other platforms using the appropriate software tools. Other WAP gateways are also available, such as that supplied by wap3gx, but the source code is not available and a charge is made for its use.

In order to complete the network interfaces in a minimal system, an SMSC interface is required. This is to supply the necessary notification and acknowledgement messages required in an MMS message flow. The Kannel Foundation also supplies an SMSC gateway, which supports protocols like EMI, CIMD (Computer Interface to Message Distribution), SMPP (Short Message Peer-to-Peer Protocol) and SEMA that are used by SMSCs. These protocols define how SMS messages are sent and retrieved from SMSC, and are employed when using WAP over SMS.

In order to make an MMS system profitable it must interwork with a billing system. Generally speaking, there are no open source billing systems available: the systems installed by most operators and service providers have evolved over many years and usually consist of many separate components. Billing system integration is considered to be an art. Until recently, interfaces have been proprietary and unpublished, unlike networking protocols. So, a supplier of an MMSC would have to work with the operator to integrate with existing systems. To reduce the amount of effort required, the supplier could adopt an integration method: the easiest method would be to supply CDRs to the external billing system, using the established FTP system to perform the transfers. Tighter integration could be supplied using the OSA standard, in which a number of interfaces for charging and accounting purposes are defined. The transport mechanism between the MMSC and billing system could be CORBA or XML: both have open source implementations available that could be used to develop the billing interface.

The application software forming the actual MMS relay/server would have to be further developed, due to the lack of an open source alternative. There are a number of software tools available for this purpose, such as Java Servlets and CGI (Common Gateway Interface). The core components of Apache Web Server and Sendmail Mail Transfer Agent will be vital in building an open source MMSC. A number of further interfaces are specified using XML, which could be developed using one of a number of appropiate XML development tools along with available transport protocols such as SMTP, HTTP, etc. Visit **www.x-smiles.org** for an open source Synchronized Multimedia Integration Language (SMIL) player.

In conclusion, the adoption of open standards with software running on commodity hardware represents a golden opportunity for suppliers to get into a sector that has been traditionally dominated by a few suppliers.

5.13 NETWORK CAPACITY AND APPLICATION LIMITATIONS

MMS brings the concept of complete multimedia delivery to the palms of users a bit closer, but there are still limitations that would prevent them from obtaining, say, a complete CD's worth of music or a complete feature film. Network band width prevents a large multimedia file being sent to a hand-held device in an acceptable time frame, and storage limitations on the devices would prevent the files from being completely stored; so, applications have to be devised that optimize use of the device. Applications that make use of existing technology and minimize such limitation would be those that allow the end-user to choose the clips or highlights they require. So, instead of retrieving an entire football match end-users could select the parts of the match they want to download, or view a match report from, say, an impartial expert, the team's manager, a player, a fan, etc. Similarly users could select news clips they want to view. This theme could expand into advertising and marketing. Where it is not practical to download a film, it may be possible for the distributers to make clips available. Another use may be for video chains to make a trailer available for every film they distribute. Clips could be browsed at home or in the video shop, and the selection made and paid for using the system; it would then just be a matter of collecting the film, or getting it delivered.

5.14 CONCLUDING REMARKS

This chapter highlighted the principal network elements required to develop an MMS system. Although many problems remain the difficulties involved in integrating the system are significantly simpler than they have been in past telecommunications systems: intelligent network protocols have been traditionally complex and accompanied by equally complex encoding schemes. This has mainly been driven by previous ideas that network capacity is more important than computing power as far as signalling purposes are concerned. With the abundance of network capacity, the development of signalling has moved into an open arena, with the adoption of plain text encoding schemes, using HTTP and SMTP as transports running over TCP/IP, significantly reducing costs and development and integration times.

The concept of moving multimedia content between two end points – whether two users or a content supplier and an end-user – is a relatively simple one: the systems required to generate revenue is where the complexities lie. The concept of the MMBox has been introduced, not only to cater as a means of supplying a method for users to store content away from already overburdened devices (no matter how big a hard disk is, it is never enough to store all those MP3s available) but also to provide an extra revenue stream in the form of charging for storage. MMBox is also an advertising opportunity: subscribers could get free persistent storage if they accepted adverts (and reply to them).

CDRs at every stage have been well defined in this chapter and may be of use in the design of new billing systems. In addition, the concept of reply charging, which allows the sender of an MMS to pay for responses that are sent, an important feature for advertising and marketing purposes – has been dealt with in depth.

The conclusion that can be drawn is that the MMS's set of stan-dards permits the entry of suppliers who have not been tradition-ally associated with the telecom industry and signifies a crossover between software development associated with Internet and Web industries and that of the telecom industry, reducing costs in the process. One of the important developments is that the boundaries between the industries will become fuzzy, to the point that eventually they will be indistinguishable. Successful implementations will integrate with existing systems and be able to operate seamlessly with other networks.

Part III

Multimedia Messaging Services Today and Tomorrow

6

Multimedia Messaging Services Today

6.1 INTRODUCTION

Multimedia messaging is still in its infancy, with operators trialling services to get a feel for the full implications of offering them to subscribers. Let us now look at a selection of services currently being offered by operators around the world, as well as the tariffs associated with them.

6.1.1 Orange in the UK

Multimedia phone users are offered the latest news with pictures every day. They can be sent the latest news, celebrity gossip, sport and quirkiest stories – all with pictures:

- celebrity – latest celebrity news and pictures sent to multimedia phones daily;

- news – the top UK and world news and pictures delivered twice a day;

- sport – a daily round-up of the top sports stories with pictures;

- quirkies – a daily delivery of stories and pictures that can be sent on to friends and colleagues;

Multimedia Messaging Service   Daniel Ralph and Paul Graham
© 2004 John Wiley & Sons, Ltd   ISBN: 0-470-86116-9

- football – news, previews and match reports for the user's Premiership team.

6.1.2 Mobistar Belgium

Mobistar Belgium offer a selection of services:

- news headlines – with colour photos;
- sports – colour pictures accompanying the results of your favourite sporting events (soccer, tennis, Formula 1 and cycling);
- weather forecast – in colour and for the whole of Europe;
- finance – follow the progress of your investments on colour graphs every day.

The take-up of Mobistar's services is expected to be slow due to laws in Belgium that prevent handset subsidies (one of the main methods operators use to kick-start the take-up of services).

6.1.3 O₂ in the UK

mmO$_2$ offers a range of Multimedia Messaging Services (MMSs) across all its European territories, including the UK, the Netherlands and Germany. O$_2$ customers with MMS-enabled handsets will be able to compose their own multimedia message (MM), comprising a mix of personal photographs, pictures, sound recordings as well as text and send it to another mobile or email address. They will also be able to access the O$_2$ portal to download images from a photo album and image bank. A range of multimedia text alerts covering sport, entertainment, gossip, news, travel and weather is also offered. The service is marketed to consumers and business users by emphasizing its compelling content and applications, ease of use, competitive pricing and comprehensive instore customer support. mmO$_2$ plans to remove some of the charging complexity of other competitive offerings by introducing two simple tariff choices: customers will be able to select either a pay-as-you-use service, which costs from around 30 pence per message, or choose from a range of bundled services that give varying numbers of messages for a flat rate fee. mmO$_2$ is working with established media brands and innovative new media start-ups to attract a wide range of MMS content. It has already signed deals with Arsenal and Bayer Leverkusen, Hallmark, Ringtones Online and Warner Bros (via Motorola). Through its rapid development centre, Source O$_2$ it will also continue to work closely with third-party application developers on MMS. To date, there are more than 7,000 registered members of Source O2.

mmO$_2$ believes that the appeal of MMS services and devices will drive the uptake of GPRS (General Packet Radio Service), medium-speed, mobile data connections, resulting in important new revenue streams for both operators and content providers. Kent Thexton, Chief Marketing and Data Officer, said: "We believe that MMS has mass appeal – from teenagers and parents exchanging photographs and experiences to business users sharing product information. Customers will be able to send a message incorporating unique and spontaneous photos and sound to share with friends and family for little more than the cost of a first class stamp. MMS is a natural evolution of the strong SMS [Short Message Service] trend."

6.1.4 M1 of Singapore

Singapore's second largest mobile operator, M1, has extended its free MMS promotion to customers until 1 January 2003. M1 was the first to launch MMS in Singapore (on 3 August 2002) and its customers have been enjoying free MMS ever since. MMS will cost $0.30 (30KB or less) and $1.20 (31KB or more), but will form part of the free bundled SMSs package that all M1 customers are entitled to. Those who are entitled to 300 free SMS per month can enjoy a mix of free MMS and SMS, as long as the total cost does not exceed $15. The charges of $0.30 and $1.20 apply to intra and interoperator MMS traffic as well as to Global MMS. Inter-operator MMS was made available on 28 November 2002. Since the launch of MMS, the number of M1 customers using this service has been rising steadily and has now hit 15,000. Neil Montefiore, Chief Executive Officer (CEO) of M1, said: "We like to think in terms of customer needs and from that perspective it will make a lot of sense for us to include MMS as part of the free bundled SMS. MMS is basically an enhanced form of SMS and people who habitually use SMS are likely to adapt very well to adopting MMS. The free bundled MMS and SMS package will appeal especially to this group of customers as it gives them added flexibility and convenience in managing and customising their mobile communication needs."

6.1.5 Tecnomen and France Telecom Dominicana in the Dominican Republic

Tecnomen is to begin trials of MMS experimental services with France Telecom Dominicana, the Orange telecommunications operator for the Dominican Republic. The Tecnomen MMSC (MMS Centre) also supports prepaid functionality and allows operators to employ various billing models for MMSs.

Among the services deployed for both legacy and multimedia handset owners are: Person-to-Person (P2P) MMS, Multimedia Mailbox (MMBox) (offering routing and filtering for MMs) and permanent MM storage. France Telecom Dominicana has also deployed the Tecnomen MMSC integrated push feature, which allows operators to deploy MMSs without having to invest in a WAP PPG (Wireless Application Protocol Push Proxy Gateway). "We are very confident that Tecnomen MMSC will be a success. The trials will give France Telecom Dominicana the opportunity to fully test the multimedia messaging services against a number of handsets and billing models," says Kimmo Aura, General Manager, Tecnomen Americas. France Telecom Dominicana covers approximately 95% of the existing network within the Dominican Republic. These trials are essential to increasing our 3GSM (Third Generation Services on a GSM network) strong-hold to enable us to provide greater access to reliable and affordable telecommunications."

The trial was set up through Tecnomen Latin America, which services a number of network operators in South America and Central America. Tecnomen will begin MMS trials with Indosat IM3, a major Indonesian telecommunications operator. Available globally, the Tecnomen MMSC platform enables operators and service providers to deliver rich MMSs to highly targeted customer segments. Tecnomen MMSC also supports prepaid functionality and allows operators to employ various billing models for multimedia messaging services.

6.1.6 Persistent store provider in the UK

MMS Store (**www.MMSstore.com**) offers a persistent message storage facility that offers the following features:

- an MMS Store content mailbox charged at a flat rate;

- assistance in getting started is immediately available from its customer service and technical teams by email and telephone to all subscribers;

- Mobile Streams (part of MMS Store) is the largest independent operator of picture message services;

- MMS Store has been designed to work with MMS-compatible phones that have an email client, which means that subscribers can use MMS Store even before their network enables MMS;

- the service is available to anyone connected to any mobile network that has launched MMSs anywhere in the world;

- it operates with a wide selection of MMS mobile phones irrespective of the manufacturer.

6.1.7 ITN in the UK – an example of a Value Added Service Provider (VASP)

The UK-based news organization, ITN, offers customers access to specially formatted ITN video news bulletins that are typically of one and a half minutes' duration, updated throughout the day and cover the top five UK and international news stories. There is also a daily special bulletin that focuses on the main news story of the day, as well as breaking news stories.

Lisa Gernon, Strategy and Marketing Director at 3 (a mobile operator), said: "One of the many strengths of 3G as a medium is that it unites the convenience of mobile telephony with the immediacy and emotional impact of mobile video. ITN's news service is a very important addition to our offering, delivering the latest video pictures from around the world straight to the customer's handset." Michael Jermey, ITN's Director of Business Development, said: "Putting the full drama of television news footage on a mobile phone is an exciting first for ITN. People will be able to call up the latest pictures from ITN while they are on the move. ITN led the way in launching text news on mobile and we're extremely pleased to be the first to reach the next generation, making our news footage available to a new audience."

6.1.8 Hutchinson Whampoa's UK 3G service

Hutchinson Whampoa's 3G mobile operator in the UK is 3, which delivers the option of three price plans. Option 1 – 3ToGo – offers:

- no monthly line rental;

- allowed the subscriber to send emails and explore news reports, sports news and fixture listings, share prices, Lotto results and more free;

- video downloads from 50p each, depending on content, including FA Premier League action, cartoons, comedy and more;

- text messages charged at 10p each, picture messages at 25p and video messages at 50p;

- up to £10 worth of video downloads free each month for your first three months;

- buy two handsets and get £100 off the total price, plus a 20% discount off the price of your video calls.

Option 2 – Kit On 3 – is a "bundle" of voice calls, video calls, video downloads and interactive content, including:

- 250 text messages, 50 downloads, 60 picture messages and 40 video messages;

- allowed the subscriber to send emails and explore news reports, sports news and fixture listings, share prices, Lotto results and more free;

- fixed monthly cost of £59.99 for a minimum 12-month contract.

Option 3 – Caboodle On 3 – doubles the Kit On 3 "bundle'" for less than double the price:

- 500 text messages, 100 downloads, 120 picture messages and 80 video messages;

- allowed the subscriber to send emails and explore news reports, sports news and fixture listings, share prices, Lotto results and more free until 30 June 2003;

- fixed monthly cost of £99.99 for a minimum 12-month contract.

6.1.9 Australian mobile operators

Australia's Telstra offers services that enable customers to send personalized postcards on their mobile phone. "Customers can tailor their messages with colour images and audio to portray a moment in time, share emotions and highlight personal triumphs," said Ms Keele, Telstra's CEO. "One of the advantages of this service is that messages can also be sent to a non-MMS handset where an SMS notification will be sent to the recipient advising them of how they can view the message on Telstra.com."

Rival network, Optus, has launched its MMS, which runs on Nokia hardware. Beyond the launch of MMS, Nokia and Optus are also actively cooperating on future MMS application development though their joint FutureLab project. Richard Kitts, Director of Operations at Nokia Networks, said, "We are providing Optus with an MMS solution for content access as well as person-to-person mobile messaging including from handset to email, email to handset or between applications and handsets. This will help ensure that the user experience is simple and fun." Allen Lew, Managing Director at Optus Mobile, said, "... we will continue to provide Australians with access to the latest mobile technology and more importantly applications and services which are useful and relevant in their daily lives."

Unlike the popular SMS (text messaging), MMS messages can include photos, animation, speech and audio files downloaded to the mobile phone from WAP sites. Globe's initial set of MMS service applications include: Traffic-Cam, which allows subscribers to view real-time pictures of actual road traffic; Animated-Messages, which allows the sending and receiving of Disney pictures and melodies; and Photo-Messages, which allows transmission of full colour pictures and sound. Meanwhile, Optus' new MMS content includes:

- news – get the latest general, business, entertainment and sports news from AAP featuring a photograph with each main story;

- SurfCam – see a picture of the latest surfing conditions at one of 20 beaches around Australia including Manly, Palm Beach, Cronulla, Portsea, Burleigh Heads and Surfers' Paradise (sourced from one of Australia's leading surf reports – Coastal Watch).

To get the new colour MMS content, Optus customers simply go to the Optus website at **www.optus.com.au/mms** and register for the content they want to receive, which is then sent to them at the specific time they request. Each message costs 75 cents with the exception of MMS mobile cards, which cost $2.95.

6.1.10 MMS roaming between Singapore, Australia and the Philippines

SingTel, Optus, Globe, AIS and Telkomsel have also launched MMSs. MMS roaming between Singapore, Australia and the Philippines is already available.

Nokia is supporting Optus's launch of Australia's first commercial MMS by providing an end-to-end MMS network solution. Launched at the end of August 2002, MMS is the next generation of mobile messaging and allows users to send and receive colour pictures, photographs and even audio messages, signalling a shift from text-based SMSs to multimedia communication via mobile phone handsets.

MMS has up to the moment only received a lukewarm reception by Australian mobile users, but is looking forward to a welcome boost following news that an intercarrier agreement for mobile phone picture messaging is imminent. Telstra, Vodafone and Optus all began offering MMSs, charging about 75c per message.

Both telcos and handset manufacturers are hoping MMS will prove lucrative in the short term, as it requires little additional infrastructure on existing mobile networks compared with 2.5G (Second and a half Generation) and 3G technologies. However, some analysts have questioned whether the obvious youth target market will take to MMS, as the new handsets that support picture messaging mostly retail for more than $1,000.

Text messages, or SMSs, have proved a runaway success in many countries (including Australia), with millions of messages sent every month for 20 to 25c. Originally, SMSs could only be sent between users on the same network. The popularity of text messages exploded after intercarrier messaging was introduced in April 2000, with Optus reporting that SMS traffic increased 1,000% in the first 12 months.

6.1.11 MMS in Sweden

Competition is heating up in Sweden's MMS market with the newly merged telco Teliasonera setting a lower MMS price than rival Vodafone. Since the launch of MMS in Sweden, Vodafone has witnessed a significant uptake of its service – Vodafone Live! The operator has set the price per MMS at SEK4.80 (£0.52) from 1 February 2003. In comparison, Teliasonera has set the price for its service at SEK3.80 (£0.41) from the same date onward. Another operator Tele2 kept its service free the longest (until 1 March 2003) but has yet to confirm its pricing strategy. Due to the increased competition there was an MMS price war at the beginning of 2003 among Swedish operators.

Teliasonera's launch of MMS now means that all three Swedish operators have launched the service. In line with its competitors' strategies, Telia Mobile also offered a free introductory period up until 15 February 2003, after which the cost was SEK3.80 per message, which is just SEK1 less than Vodafone Sweden is charging. Tele2 launched its MMS in late November 2002 but, unlike its rivals, has yet to announce details regarding pricing.

Devine Kofiloto (Research Analysts) European Team tracks interoperability MMS agreements in Europe, while EMC World Cellular Data Metrics tracks MMSC contract award, launch status and tariffs (among many other mobile data indicators including MMS traffic) for key European operators. Consequently, Sweden could end up with a large number of MMS users, but predicted profit will be much lower in the first phase.

InfoSpace powers a variety of branded MMS-based applications on AT&T Wireless mMode(sm) service. mMode subscribers can access InfoSpace-powered multimedia data services, such as full weather alerts including forecasts, satellite images and weather maps from **AccuWeather.com**, icon-based horoscopes from iVillage's **Astrology.com**, and finance charts and information from Silicon Investor(R), and more.

6.1.12   TIM and the Walt Disney Internet Group in Italy

TIM and the Walt Disney Internet Group have entered into a content arrangement that will bring Disney Mobile-branded wireless content to Italian consumers. As part of the agreement, TIM will be the first distributor in Italy for a range of Disney's wireless content. Disney Mobile's content offering to TIM customers is a complete line of products including ringtones, logos, screensavers, wallpapers, picture messages and games. It will be based on Disney's line-up of classic characters, including Mickey Mouse, Minnie Mouse and Donald Duck, as well as on characters from modern classics such as *Beauty and the Beast* and *The Lion King*, and more recent blockbusters such as *Monsters & Co.* and *Lilo & Stitch*. The content will be regularly updated to include characters from future Disney movie releases.

The Disney Mobile service can be accessed through the TIM Web and WAP portals. On the i-tim website (**www.tim.it**) TIM has created a virtual area entirely dedicated to Disney, taking the consumer through previews of all the content available. TIM and Disney are pioneering partners at the forefront of MMS technology. The combination of moving characters, music and personalized messaging that MMS enables really brings the Disney magic to life on a phone's display and sets a new standard in Europe for wireless multimedia entertainment. With this launch TIM and Disney are giving a powerful demonstration of their common commitment to new communication technologies. Over time Disney's content offering on the TIM network will be expanded to take advantage of the opportunities arising from Java and videostreaming applications.

6.1.13   Cellcom in Israel

MMS users can send and receive MMs with rich image, audio, video and text elements. In addition to P2P, Comverse's MMSC

Version 2.0's application-to-person messaging capabilities pave the way for visually compelling content applications that can further drive MMS usage. Quick deployment of new applications and services is made possible by the Extensible Messaging Framework (XMF), a flexible MMSC architecture that enables easy modification of system behaviour to accommodate operator requirements and preferences. "Comverse's solution is in line with Cellcom's strategic plans to launch advanced multi-media services, based on its GPRS network," said Dr Yitzhak Peterburg, CEO at Cellcom. "It has the flexibility that meets our specific and demanding requirements for advanced media." Zeev Bregman, CEO at Comverse, noted, "Our customers are our best sales representatives. While it is important that our technologically and functionally advanced MMSC with extensible Messaging Framework [XMF] and advanced transcoding capabilities exceeds standards and transcends basic requirements, customer satisfaction proves to be equally important and one of our most valuable assets."

Important MMSC elements to be deployed at Cellcom include: the MMBox, a multimedia network-based personal storage box; the Multimedia Album, which makes attractive rich media elements easily available for messaging; and enhanced OA&M (Operations, Administration and Maintenance) capabilities.

6.1.14   Sonofon in Denmark

Danish telco Sonofon has launched its MMS provision in Denmark through a turnkey solution delivered by mobile application infrastructure provider End2End, and Openwave Systems Inc., a mobile software developer. The turnkey MMS solution is based on Openwave's MMSC relay, message store and directory, and is delivered as a turnkey-managed service by End2End, a managed service provider for mobile data solutions in Europe. Sonofon's customers have been able to send photo messages as well as create and distribute other MMs since 20 December 2002. To do this, they had to use a composer available on their website (**www.sonofon.dk**), using pictures from a personal gallery service alongside MMS animations delivered by Iceland-based ZooM Hf.

6.1.15   Telia Mobile in Finland

Nokia and Telia Mobile Finland have agreed on the supply of a complete Nokia MMS solution, as well as formation of the Nokia Charging Centre (NCC) for state-of-the-art charging for services. The MMS system is already in use with Telia Mobile Finland's commercial MMS service. The MMS solution provided by Nokia includes the Nokia MMSC, the Nokia Multimedia Terminal Gateway for support of non-MMS phones, the Nokia WAP gateway and the Nokia Profiler Server. In addition, Nokia will provide system implementation, integration and care services to maintain the competitiveness of the MMS solution. The market-leading Nokia MMS solution allows Telia Mobile Finland customers with MMS-capable phones to exchange rich MMs containing text, images, graphics, voice and audio clips. Thanks to the Nokia Multimedia Terminal Gateway, MMs can also be sent to legacy phones that are not MMS-capable. The NCC is a comprehensive charging solution for MMS as well as for SMS, GPRS access and other data services, such as content downloading. The NCC supports charging of both prepaid and postpaid customers and meets the challenges of next generation networks and services by providing sophisticated charging based on metrics, such as data volume, content and quality of service.

6.1.16   StarHub Singapore

StarHub has launched a new TV channel targetted to get teenagers to use the service and watch the programme. Named Hub TV, Singapore's first MMS and SMS TV channel allows viewers to "actively influence the contents" on screen. Viewers can take part in interactive activities, such as polling, chatting and participating in contests using text messaging while watching TV programmes. They can even send in pictures to share with other viewers using multimedia messaging. The Hub TV screen is divided into four main sections: the video panel for TV programmes; an SMS "live" chat panel; the MMS board panel; and the contests/polls panel. On the SMS "live" chat panel, viewers can participate in an interactive chat by sending their messages, which are displayed on the TV screen. This chat is open to both StarHub Mobile and M1 customers (each SMS is charged at 30 cents). Similarly, MMS pictures and messages are displayed on Hub TV's MMS board panel. "With Hub TV, another youth-oriented service, youths no longer play the passive role and just sit in front of the TV screen. They are empowered by StarHub to actively influence the contents on Hub TV by sharing their thoughts and chatting via SMS or expressing themselves visually by sending pictures via MMS. Best of all, they can do all these activities simultaneously while watching their favorite artiste like F4," said Mr Chan Kin Hung, Senior Vice President at StarHub Mobile. "Hub TV has become a powerful two-way interactive medium for our young customers to share, express and be entertained!"

6.1.17 Beijing Mobile in China

Huawei Technologies, a global provider of network equipment and solutions, has helped Beijing Mobile, a subsidiary of China Mobile, to launch an MMSC in Beijing, which opened to traffic on 1 October 2002. The MMSC is the first commercial one in China, signalling the advent of a new multimedia communications era. For this MMSC, Huawei's multiple MMS applications include terminal-to-terminal, terminal-to-applications and applications-to-terminal. The Huawei-contracted MMSCs of China Mobile are located in Beijing and the major industrial centre of Wuhan. The advanced messaging services will enable China Mobile to deliver its service to 1.25 million customers in different provinces across China. Using Huawei's MMSC, China Mobile allows its customers to receive and send MMs containing texts, images, graphics, voice and audio clips easier and faster, at any time and anywhere.

6.1.18 Orange in Switzerland

Nokia and Orange Communications SA Switzerland have signed a three-year contract for the delivery and implementation of Nokia's MMS solution. Orange Switzerland launched its commercial service on 24 October 2002. In addition to the MMS platform, Nokia is providing services for implementation, systems integration, support and maintenance to ensure and maintain the competitiveness of the MMS solution. Nokia supported Orange

Switzerland in promoting MMS P2P messaging at Orange's cinema events in five Swiss cities in July and August 2002. Its MMS wallboard solution employed Nokia 7650 imaging phones and offered the cinema visitors the possibility to try MMS services prior to the full commercial service launch. Approximately 250 photos were captured each evening in each city, and visitors were able to view the MMs simultaneously at all five locations. The end-to-end multimedia messaging solution that Nokia supplied includes the Nokia MMSC, the Nokia Multimedia Terminal Gateway for non-MMS phones, the Nokia WAP Gateway, the Nokia Multimedia Email Gateway and the Nokia Profile Server. The Nokia MMS solution gives mobile subscribers the ability to enjoy multimedia services, such as content access and P2P mobile messaging from MMS-capable handsets to email, from email to MMS handsets or between applications and MMS handsets. The Nokia Multimedia Terminal Gateway even makes it possible for users of non-MMS-capable legacy Global System for Mobile communications (GSM) terminals to receive MMs. Legacy-terminal users receive an SMS notification and can connect to a website to view their MMS message.

6.1.19   Plus GSM in Poland

Plus GSM was the first operator in Poland to launch MMS commercially. The price is 2 PLN per MMS and the size of MMS can be up to 50KB. The service is available to all subscribers without any activation or monthly fees (the charge is just for messages sent). Plus GSM subscribers can send MMSs to mobiles and email accounts. If a subscriber does not have an MMS phone, he or she receives an SMS with information containing the name of the WWW page: **foto.plusgsm.pl** and the unique login and password to that particular MMS.

6.1.20   MMS used to enhance interactive TV

Netherlands-based interactive TV powerhouse Endemol – the developer of ratings champion *Big Brother* – has unveiled its upcoming range of programmes. The company is moving even further toward MMS and SMS-related mobile content: moving from television programming that involves mobile interactivity to mobile interactivity that involves television programming.

6.2   MOBILE PERSONAL COMMUNICATION

Although MMS is obviously used to communicate between two people, either using traditional messaging or more expressive picture messaging, there may also be instances when a person would want to send a recorded message, such as a greeting to be listened to at a certain time (i.e., birthday greetings) or instructions that a person may want to listen to several times. Another feature of the audio messaging features of MMS is that it can be used for advanced voice mail. In case the end-user is not available, then a remote server can store the message in MMS format, if the caller is calling from a land line or legacy system, or, if the phone is MMS-enabled, the caller can record the message and send it as an audio MMS message. Either method will be stored in the called person's MMBox for retrieval later.

6.3   MOBILE DATING

Mobile dating agencies have the potential of being highly lucrative services: Internet-based online agencies have proved to be popular, and the convenience and privacy of MMS guarantee making it a very popular service. This different type of message may warrant a staggered approach: initial contact could be made using text messages stored in a MMBox; the next phase could involve audio communication; finally followed by pictures and video messages when both parties are certain they wish to maintain this level of contact, thus preventing malicious use of the service (**http://mobile.match. com/**).

6.4   MOBILE MARKETING

The graphical potential of MMS is a good complement to branding and advertising campaigns. A user could opt into a free information service from a movie studio or a cinema chain. At the launch of certain movies, the studio could send an MMS including pictures and a description of the movie, and eventually even a short trailer. The users could then exchange the message with friends to invite them to go to the cinema. By distributing MMS messages, the movie studio could then increase interest and awareness of their products. However, the market for wireless advertising should remain strictly opt-in. Unsollicited messages are considered as spam by users, to the extent that the UK now has laws prohibiting email spam (Engdegard, 2002; Patel, 2003).

6.4.1 Diageo and Genie in SMS marketing deal

Diageo, the food and drink group, has formed a partnership with Genie, the wireless Internet division of UK carrier mmO$_2$. The two parties will deliver SMS campaigns across their databases. Diageo's brands include Jack Daniel's and Smirnoff. Through its subsidiary, Nightfly, it has built a database of more than 100,000 young British subscribers to its SMS information service, centred on its drinks. It has achieved this by recruiting in pubs, clubs and bars. Advertising provides a means for subscribers to use a service with no charge as long as they are prepared to view an advert or send an MMS to another subscriber with an advert attached to it. Location-based services coupled with advertising capabilities provide a powerful marketing tool, special offers may be advertised on a local basis.

6.4.2 Short-term opportunity: target MMS-curious consumers on the Web

Even though the conversion of mainstream consumers to active MMS users will take time, consumers' rapidly growing awareness and curiosity about multimedia messaging will provide marketers with short-term opportunities outside the mobile channel. Consumers are already starting to use Web-to-Web and Web-to-mobile MMS applications. Marketers hoping to capitalize on the novelty of multimedia messaging should place their brands using Web-based MMS applications offered by portals, communities and mobile operators. Mobile industry giants Nokia and Vodafone are among the first to have done so. The former sponsors an MMS learning section on Lycos in several markets, while the latter sponsors an MMS composition page in Germany. Coca-Cola has gone one step further in Austria by creating its own MMS application, with which consumers can buy animated greetings when visiting Coke's site.

6.4.3 Mobile-centric campaigns hold greatest promise over the long term

Marketers should shift their focus from Web to mobile when active MMS users exceed the critical mass of 10%. In the interim, mobile-focused MMS campaigns, such as Hasbro's three-slide promotion of the Monopoly game, will be foremost in generating public relations effects. In the long run, mobile-centric campaigns promise to become more valuable for advertisers. The novelty of Web-based MMSs will diminish, largely because there are well-established means of PC Internet communications (e.g., email and instant messaging). A wireless campaign has the advantage of location and behavioural targeting, fast message consumption and greater screen dominance. Nevertheless, to optimize the effects of wireless messaging campaigns, marketers should combine MMS, SMS, Web and traditional channels.

MindMatics offers a unique portfolio of interactive advertising solutions based on permitted marketing and loyalty programmes. With MindMatics' concept the user actively signs up for sponsored services with his personal profile and is always in control of what type, when and how much advertising he is exposed to www.mindmatics.co.uk).

6.5 MOBILE INFORMATION SERVICES

Technology has fed the human addiction to being kept up to date with the latest information, whether it is current affairs, sports results, stock information or show business gossip. MMS provides another method to deliver instant information to feed this desire. However, not only do people want to be informed they want to share information with others as well. With MMS, information services can be delivered that are considerably more informative than those already available over SMS, particularly regarding:

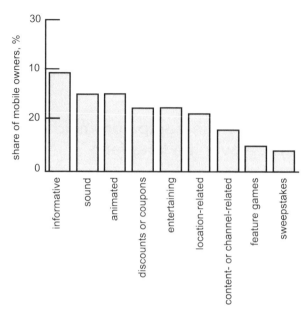

Figure 6.1 Types of MMS ads mobile owners believe will grab their attention
Source: Jupiter Research

- sport;

- finance;

- general news.

For instance, a football service might include some action on the pitch, an audio message and some text to describe the action and a financial service might include a graph showing the movements of an index or the value of a share during the day.

The key to information services is to provide them in a format that is concise and easy to view. Too much information that is hard to digest will put subscribers off using them; alternatively, if the information is too sparse, the subscribers will go to other information sources, such as the Internet, teletext, journals and newspapers. There must also be the possibility of linking to other information sources: if subscribers receive a football result and some commentary, then they may also want to view the associated league table, or to view archived material of previous matches. This leads to the "surfing" effect, which the Internet created, but with the added advantage that once a piece of information is found it can be sent on to somebody else, with the added revenue implications for operators. Although this effect is currently not possible with SMS (i.e., receiving one SMS does not lead on to numerous other messages), there is such a thing as SMS "surfing": whether the actual message is sent to another user or just a link, it will always result in a message download and thus some kind of payment, either for the message or for the volume of data downloaded. One application, Namidensetsu (which means "surfing information"), for example, became an early and long-lasting hit (Figure 6.2). It provides wave and weather information about 200 Japanese and 40 international surf spots. In addition, it offers instructional information, surf shop and tour information, advice for bodyboarders and a service for downloading popular surf brand logos. A similar site is available for fishing. Another is Popteen, an information and communication site for girls and young women in their teens and early 20s. It offers fashion information, astrology and information about celebrities.

CYBIRD has built wireless online communities for cat lovers and office workers. As part of its business solutions group, CYBIRD recently joined with Kodak in Japan, forming a partnership that combines Kodak's digital lab services with mobile Internet technologies and CYBIRD's wireless expertise. The new alliance will enable wireless users to take advantage of a variety of new

Figure 6.2 Surf information (screen graphic)
Source: **www.cybird.co.jp**

digital features. Specifically, consumers will be able to use their mobile phones to check the status of orders submitted for developing and printing conventional camera film, peruse online photo albums and place print orders directly with Kodak digital service outlets. The venture will also take advantage of CYBIRD's existing Prinet application, which enables users to convert their favourite photos to wallpaper for their wireless wait mode screen, as well as to send them to i-mode, J-Sky, ezweb, and feelH mobile phones in Japan. Pictures converted to the size of mobile phone screens may be sent through Yahoo!Photos from personal computers, through mobile phones with digital camera functions or through Print Club terminals at the push of a button.

6.6 MOBILE ENTERTAINMENT

As an aid to mobile game playing, Nokia have come up with N-Gage. Basically, it is a hand-held gaming console, with high-quality mobile phone features, and a digital music player all in one. It has been designed specifically to encourage mobile gaming, initially via a GPRS network. It is envisaged that it will be the first in a line of new generation handset devices that will exploit the capabilities of high-bandwidth mobile networks far beyond basic voice and text messaging (Bond, 2002). It will support SMS, MMS, email and smart messages, and will have a built-in RealOnePlayer for viewing video clips. We will also see game console companies getting into the mobile handset space: either with add-ons for existing devices, such as Game Boy Advance, or with totally new devices.

6.6.1 Interactive "pick-a-path" video

Devil's Night is a "pick-a-path" adventure about a group of girl-friends who unearth the evil motives of a pre-Halloween party – a discovery that may spell their doom (**http://www.fluition.com/gallery/main.shtml**). One of the services currently being considered for offer by several service providers is the "pick-a-path" video, which gives the user the ability to determine the outcome of a video film or story: at several points in the video, the viewer can be requested to determine in which direction the story should go. This theme could be extended to allow a group of friends to view the same story with the outcome determined: by any one of them, or by taking a vote among the group, or by taking turns to pick the next scene. Whichever option is selected will result in the program sending vast numbers of messages around the system. The final story could be stored for future access, or sent to somebody else in its entirety. The technology to enable these types of services can be provided by VAS extensions.

6.6.2 Interactive music selection

Aporia Records (**Aporia-records.com**) makes use of Fluition CD to showcase new bands. The player is embedded into the Aporia website to create a seamless experience for the viewer. Another service feature of MMS is the ability to download music on demand. Although we have looked at this in other sections, we have not considered interactive music (an MMS jukebox), where music can be listened to on a one-off basis. This obviously requires Digital Rights Management (DRM) systems to be in place, but users may decide that they will not listen to a piece of music enough to warrant purchasing it, but they would be prepared to pay a token fee to download it and listen to it on a one-off basis.

6.6.3 Collectable cards

Cards or stickers are frequently collected by young users in many markets. The cards typically have on one side a photo of a sports-person, a movie character or a car, etc. and on the other descriptive text. MMS can deliver the same service, with users paying to receive a random set of cards (messages). In addition to the revenues generated by the delivery of each card, the service also generates traffic, by allowing users to exchange cards among themselves. Service providers will have to consider the trade-off between card exchange (traffic) and card delivery (payment for content). Too much exchange and most users will receive cards for free, while rules on content sharing that are too strict will put users off and lessen or even remove the fun of trading. Hence, the service provider needs to implement a DRM system than can accommodate both. For instance, when forwarding a card, the original copy on the terminal could be deleted (a practice called "forward and delete"), thus providing a real exchange of cards and not a simple duplication.

6.6.4 Celebration cards

Prepackaged MMs, such as birthday cards similar to the gift cards sold in shops could again be popular and a further source of revenue.

6.6.5 Adult services

Adult services will play an important part in mobile data services, as they already do on the Internet. However, pornography, erotic content and dating services are difficult subjects for many operators, who are rightly wary of associating their brands with this content, because it can be offensive to many. This market is already one of the most successful services on the fixed Internet. Even though mobile terminals have smaller displays, they are personal, portable and hideable – perfect for discreet usage of such content. Not all operators will market erotic content: some will partner with specialist content providers, as Virgin is considering doing with *Playboy*; others will simply enjoy the revenues from the traffic generated by other content providers. Adult services are one of the main drivers for prepaid MMS services. Subscribers can make a payment to an anonymous prepaid account and access the services with no record being made against their name.

6.6.6 Comics, jokes and icons

"Joke of the day" is already a common service in text form over SMS. With MMS, comics can also be delivered and additional traffic thus generated. Eyematic provides cartoon icons to represent the user's personality in a virtual mobile world (see Figure 6.3).

6.6.7 Horoscopes

There are already successful horoscope services via SMS, but MMS will make horoscopes look and sound much better, though the basic functionality of the service will remain the same.

Figure 6.3 Example of an MMS chat icon (screen graphic)
Source: **www.eyematic.com**

6.6.8 Quizzes and competitions

With MMS, service providers can deliver a quiz in a more compelling way than by simple texting. Quizzes and competitions can be sponsored by an advertiser or paid for by the user, who can respond to the competition by returning the messages completed with answers or by sending a simple text message.

6.6.9 MMS and the young person market

As happened with SMS, many believe the young person market will make up the core of early adopters of MMS. Business users are also expected to be a key target group. Chris Lennartz, Marketing Manager for CMG Wireless Data Solutions, says research has shown that the young person market has the spending power to purchase expensive MMS phones and services and that young people could consider the new service as a means of expressing their individuality, maturity or belonging. However, Dion Price, CEO of Mobile Streams, says operators are ignoring the following MMS young person market dynamics: MMS as a service compared with SMS is expensive, making it prohibitive to use; MMS handsets are high-end devices with high-end features like colour screens and built-in cameras and as a result could be prohibitively expensive to the youth market; and prepay MMS will only come later on, so the millions of young users who are not eligible for a contract phone will be cut out of the market. Price goes on to say that one potentially lucrative MMS application "chat", where users can share real-time pictures because of the advent of handsets with built-in video cameras. "Value-added services available in MMS make it an attractive medium for people across all markets. Mobile phone users in the future are more likely to demand improved features in their handsets like being able to download and store colored pictures and exchange audio and video clips," she added.

6.7 MOBILE BUSINESS APPLICATIONS AND SERVICES

Multimedia messaging tends to be associated with entertainment and fun activities, and is essentially a mass market opportunity. Nevertheless, there are potential niche applications that could be supported for business usage. One obvious business use is the property market: estate agents could set up as VASPs and send details and pictures of properties to interested subscribers. There is also a potential use in law enforcement, where a crime is captured using a camera in a mobile handset and sent to the police as evidence (the legalities of which still need to be sorted). There has been a case of criminals being caught using camera phone evidence. The nature of the MMS also serves as a deterrent, because the message can be sent instantaneously to a police station, preventing a criminal from destroying the evidence.

As mobile handsets are carried by people the whole time, there is an increased likelihood that somebody at an accident scene will be in possession of one, thus recording the scene for insurance cost adjusting purposes. An example of a service that goes well beyond basic use, is the video alarm service from Opensuger (a France Telecom-owned company) that allows users to observe live images from remote cameras and to be notified by SMS, MMS or email when motion is detected. Possibilities for the service extend much further than alarm surveillance: it could be used for child monitoring, for example.

6.7.1 MMS law enforcement

In Italy MMS has been used by the police to capture two criminals, recording what is thought to be the world's first conviction thanks to a tip-off using an image sent by a mobile picture phone. The two thieves were jailed for six months after a tobacconist became suspicious as they loitered outside his shop. He snapped the two men with his new Nokia picture phone and promptly sent it to local police, who quickly realized that the two men, both in their early 20s, were wanted in connection with a series of raids and raced round to arrest them. The *Daily Mirror* quotes Mario Pietrantozzi of Rome police as saying: "We believe this to be the first time in the world a picture text message has been used to help catch criminals and secure conviction." This novel use of MMS could well spark its widespread use in this manner, potentially acting as a deterrent, as criminals will not be able to determine who has a picture phone and who does not (**http://www.theregister.co.uk/content/59/29346.html**).

6.8 CONCLUDING REMARKS

The multimedia services of today are many and varied, but only touch on the multitude of possibilities imaginable. The existing range of services from plain P2P messages, news information, weather reports to complex information services is extensive, but most operators are still at the beginning of MMS deployment and trials. The average user is considered to be the typical early adopter of new technology: either technological enthusiasts or businessman who are in a position to be able to demand the use of such services for their job. So, services will be based around the profiles associated with these groups: typically, these are news, financial information, weather reports, sport, email-type services and services relating to technology. The second major group to be targetted will be teenagers, although the cost of handsets and services currently puts MMS beyond most in this group and the available services will reflect this. Based on past experience of the adoption of the technology, the average person will take up MMS long after it is introduced (SMS is a good example). Lead times can be reduced if "must have" services are introduced at affordable prices or if people believe they are missing out by not using the service. One method of doing this would be to target television programmes that people have become addicted to, then offer them more information that is only available using MMS. This may simply be extra information, extra clips from reality TV-type programmes or extra scenes from soap operas. The Dutch company

TV powerhouse Endemol (*Big Brother*) is attempting this, as it sees this as the way forward for its current format. But the current killer app is still the ability of an individual to take an impromptu picture and send it to a number of friends, making him or her the number one supplier of content.

7

Future Recommendations

7.1 INTRODUCTION

Multimedia messaging is the beginning of a whole host of services offering increased visual and audio experiences. Technological advances will increase the reach and usefulness of Multimedia Messaging Services (MMSs), as well as providing the foundations for further service innovations. An example of this is the concept of the Multimedia Broadcast/Multicast Service (MBMS), which is simply to permit content to be broadcast to numerous users in an efficient manner. MBMS will be explored further in this chapter. Even fixed line operators can join the party: the F-MMS forum (**www.f-mms.org**), initiated in December 2002, is aiming to deliver the MMS to PSTN (Public Switched Telephone Network)/DECT (Digital Enhancement Cordless Telecommunications) phones. A fixed line Short Message Service (SMS) service was standardized by the European Telecom Standards Institute (ETSI: ES 201-912) and operates in several European countries, where it is beginning to grow into a new market.

For operators to invest in extra services, there have to be suitable billing systems to cope with the advanced tariffing features that come with service developments. So, charging systems will develop in parallel with services and be interdependent.

We will also look at the concept of video messaging. Although it is still under development and users are unable to create and send their own video content, this aspect of MMS will mature in the near

Multimedia Messaging Service   Daniel Ralph and Paul Graham
© 2004 John Wiley & Sons, Ltd   ISBN: 0-470-86116-9

future, offering further opportunities for operators to develop revenues and increase the possibilities open to the creators of content.

Finally, it is hoped that this chapter will aid creative readers in developing their own radical, new MMSs and technologies.

7.2   MOBILE MESSAGING EVOLUTION AND MIGRATION

Not surprisingly, many issues still need to be addressed before MMS can be declared a success. The growth of the MMS service penetration rate among mobile subscribers must remain steady throughout 2003 and into 2004 for the momentum behind MMS to become unstoppable. How can this goal be achieved? There are a number of ways:

- by allowing MMS to roam increasingly between countries and single operators so that people can communicate across "borders";

- by improving handset usability and introducing compelling phones with larger screens, better sound quality and more vivid colours;

- by developing ways to deliver content to subscribers that is always in the right format – be it MMS, WAP (Wireless Application Protocol), SMS, smart messaging, EMS (Enhanced Message Service) or WAP-PUSH; and

- by using innovative promotions and pricing schemes for both handsets and services.

Market research on expected MMS uptake suggests that applications and content will play an absolutely vital role, especially in the early phases of the market. It is true that early MMS adopters have a very limited number of people they can communicate with. But the alternative of sending messages to people's email boxes lacks the instant sharing of a feeling or moment that is so essential and unique about the appeal of MMS. The MMS standard is much improved in terms of application connectivity than SMS ever was or is. The MM7 interface is where applications and content are integrated technically into MMS networks. MM7 represents a step in the right direction in establishing common standards in the currently way-too-complex mobile infrastructure arena. In addition to network connectivity, however, there are at least another three vital issues to resolve before we can safely expect to run a flourishing MMS content business:

1. *Money*   Any content business needs to be able to flexibly define business models and billing schemes that handle practically any conceivable service and usage scenario. Of course, there are mobile originated and mobile-terminated billing schemes, but they are not good enough. Some popular messaging services are session-based; so billing data generation must be able to identify where it all actually started and ended to define the correct price for the session. And then there are the "send to a friend" or "sponsored content" types of services, where the billing ticket is neither generated to the recipient nor the sender of the content, but to someone else.

2. *Professional management of an ever-increasing and constantly changing portfolio of applications and content providers*   New applications should be introduced quickly to maintain the freshness of content and yet be under a proper Quality Assurance (QA) control process, to avoid messing up the whole messaging network. These seemingly contradictory requirements are achievable when applying an automated and predefined application introduction process. Ideally, the content aggregator can define the service level and content integration process for each of its content partners and then let these partners run the application process themselves through extranet access to the messaging provisioning system. This considerably lessens the content aggregators' workload in introducing new content and, theoretically, gives them infinite scalability of the content business. The remaining essentials are the ability to define and impose Service Level Agreements (SLAs) for content providers and to manage the messaging traffic application by application along with throttling capabilities based on the network capacity and resources available.

3. *The nature of the business and how to get ready for it*   Setting up a successful MMS messaging application business is not only about MMS content – far from it: the fast-growing SMS content business needs a good deal of business support to scale further.

This is also true of the fast-growing number of WAP-PUSH-based services (especially for rich content download over WAP), the wildly popular ringtone and icon services running on SMS, the not-so-popular EMS services and Europe's iMode messaging services. It simply makes investment sense and improves business/system control to establish a single application interface for all content types (providers), run a common billing data generator for all messaging transactions, route and throttle all traffic through a single engine and supply content types through a single system that knows nothing of the content format or required bearer: this is simply about getting more messaging bang for the middleware buck.

MMS does not embody all messaging applications: it is simply a subset of the multimedia messaging world. There are competing application platforms that should be considered as valid options when developing applications (e.g., WAP, Java and Brew). Nevertheless, some of these competing platforms can also be used as a complement for MMS solutions and vice versa.

7.2.1   Types of multimedia messaging

Most messaging applications can be categorized in three groups, according to the parties that are initiating and receiving the message:

- Person-to-Person (P2P) – personal communication, such as texting or multimedia messaging;

- Machine-to-Person (M2P) – used to deliver reminders, information and entertainment services;

- Person-to-Machine (P2M) – mostly used as part of quiz, content and voting services.

In addition, it is possible to subdivide the M2P category even further. Depending on how the delivery of the content is arranged: the service provider may automatically deliver the messages (push) or the user may request each single message (pull). Push services include:

- subscriptions – when the user subscribes to a regular delivery service (such as financial market values at the end of the day or a daily joke);

- event-driven messages – when the user presets the service to receive updates (such as the goal-by-goal account of a particular football match or when the share value of a company exceeds a certain figure).

Pull services can be:

- one-off content delivery – when the user requests information (such as a request for the latest cinema showing or the nearest restaurant);

- one-off content download – the user requests content (such as ringtones, screensavers or wallpaper to download for his or her mobile phone).

Messaging services are a huge and growing source of revenue for operators and others. But, it should be noted that there is a major difference between revenue and profitability, and it is important for those looking to benefit from this sector to ensure that they have the proper infrastructure in place to turn income streams into good profit opportunities.

Migrating from SMS to MMS is essentially straightforward, but the steps involved need to be considered properly and planned and executed well. There are very different usage situations and behaviours and, as with any product portfolio, it is necessary to take these different elements into account and to make sure that potential revenue is not lost needlessly. Messaging is king and the key Average Revenue Per User (ARPU) generators will be those personalization services and P2P messaging that let users communicate through private pictures and self-created content, create opportunities for communities to share multimedia messages (MMs), give freedom to third parties with regard to provisioning and billing, and attract the large market of subscribers who still use smart messaging or EMS devices.

7.2.2 Market barriers

The transition to the 2.5G (Second and a half Generation) high-bandwidth infrastructure that will form the wireless transport backbone for MMS has taken longer than expected. Higher band-width is required to support the ability to send and receive rich multimedia files to and from mobile handsets without a noticeable slowdown in service. MMS will be a leading traffic generator for the expanding General Packet Radio Service (GPRS) networks, but the deployment and adoption of this 2.5G technology will occur at different rates throughout the world, the impacts of which will only be seen when carriers can roll out MMS services to consumers.

Roaming agreements among network operators have yet to be resolved for MMS. While such roaming agreements are not neces-sary for users to receive MMS content services from their carrier, they are necessary if the market is to see the explosion in P2P messaging traffic experienced with SMS roaming agreements.

MMS transaction pricing has yet to become established, as it varies considerably across operators. An MMS message will prob-ably be slightly more costly during its initial roll-out phase compared with an SMS message. Industry analysts predict an MMS message will cost 2.5 to 3 times that of an SMS message, thus ranging anywhere from \$0.30 to \$0.48 per transaction (€0.34–0.54) for P2P MMs. Early pricing examples confirm these estimates: D2 Vodafone (Germany) charge €0.39 (\$0.38) per MMS message up to 30KB in size; Sonera (Finland) charge €0.59 (\$0.57) for messages up to 100KB; and Telenor (Norway) charge 10 NOK (about \$1.28) per P2P MMS message with no published limit on message size.

One of the author's first mass MMS experiences was at a Red Hot Chili Peppers concert in March 2003. Numerous concertgoers were using their picture phones to send images of the performance. Despite being a relatively new technology, it was clear that people were using the service. However, a limiting factor is network cover-age: suppose this was scaled up resulting in a greater subscriber base and more people able to send concert pictures, then the network would arguably not be able to cope. There is also the problem of MMS bootlegging: a remedy would be to limit MMS use at the event, which would not be popular, or to charge a fee for permission to send MMSs from the venue. There are technological solutions to both.

7.3 INSTANT MESSAGING SERVICE

Ericsson, Motorola and Nokia formed Wireless Village, the Mobile Instant Messaging and Presence Services (IMPS) initiative (OMA, 2002), in April 2001 to define and promote a set of universal specifications for mobile instant messaging and presence services. The specifications will be used for exchanging messages and presence information between mobile devices, mobile services and Internet-based instant messaging services – all fully interoperable and leveraging existing Web technologies. Through its supporters, the Wireless Village initiative aims to build a vibrant community of end-users and global business partners where Internet and wireless domains converge. The IEEE Industry Standards and Technology Organization (IEEE-ISTO) provides day-to-day administrative support to the Wireless Village initiative. The Wireless Village interoperability framework includes the Wireless Village system architecture and an open protocol suite at the IMPS application level to provide interoperable mobile IMPS services among workstations, network application servers and mobile information appliances, such as mobile handsets, hand-held computers, PDAs (Personal Digital Assistants) and other mobile devices. The Wireless Village system architecture, as outlined above, describes the IMPS system and its relation to mobile networking and the Internet. This is a client–server-based system, where the server is the IMPS server and the clients can be either mobile terminals or other services/applications, or fixed PC clients. For interoperability, the IMPS servers and gateways are connected by a Server-to-Server Protocol. The architecture gives implementers more choices between individual Wireless Village servers or gateways and as a bonus the Wireless Village brand and technology. The Wireless Village server is the central point in this system and is composed of four application service elements that are accessible via the service access point:

- presence service element;

- instant messaging service element;

- group service element;

- content service element.

The Wireless Village client consists of an embedded client and a command line interface client, and communicates with the Wireless Village server to accomplish IMPS features and functions and to provide users with IMPS services. The Wireless Village system architecture is consistent with Third Generation Partnership Project (3GPP) TS 22.121 Virtual Home Environment (3GPP, 2002c) and 3GPP TS 23.127 Open Service Architecture (3GPP, 2002c). The Wireless Village initiative also specifies how it will interwork with MMS standards. It is envisaged that it would be possible for an IMPS client to be able to send and receive messages to and from an MMS client.

The interoperability between Wireless Village servers and clients, and between Wireless Village servers is achieved through the Wireless Village protocol suite, which consists of the Client Server Protocol (CSP), the SSP and Command Line Protocol (CLP). The protocol stack is shown below:

- CSP is designed to provide embedded clients in mobile terminals and desktop clients with access to the Wireless Village server.

- SSP is designed to provide communication and interaction means between Wireless Village servers and the SSP gateways. It also allows Wireless Village clients to subscribe to the IMPS services provided by different servers that are distributed across the network and Wireless Village clients to communicate with existing proprietary instant messaging networks through the SSP Gateway.

- CLP is designed to provide the Wireless Village server and the Common Language Infrastructure (CLI) client with the means to communicate and interact with each other to support the IMPS services in a legacy CLI client.

The Wireless Village protocol suite runs at the application level and is compliant with Internet Engineering Task Force (IETF) RFC (Request For Comments) 2778, RFC 2779 and IMPP (Instant Messaging and Presence Protocol) CPIM (Common Profile for Instant Messaging) models. It may run independently over different transport layer and bearer protocols.

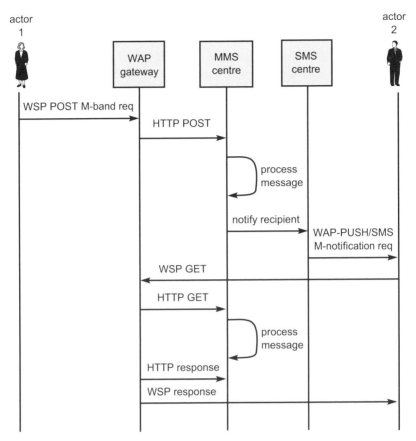

Figure 7.1 MMS usage moves through phases

MMS = Multimedia Messaging Service; WAP = Wireless Application Protocol; SMS = Short Message Service; WSP = WAP Session Protocol; HTTP = Hyper Text Transfer Protocol

7.4 MExE

Mobile Station Application Execution Environment (MExE) is a wireless protocol that is designed for incorporation in smart mobile phones (3GPP, 2002i). Its aim is to provide a comprehensive and standardized mobile phone environment for executing operator or service provider-specific applications. MExE is designed as a full application execution environment on the Mobil terminal and builds a Java Virtual Machine into the client mobile phone (Java is the "write once, run anywhere" programming language). MExE is designed to be used for provision of sophisticated intelligent customer menus and to facilitate Intelligent Network (and so-called Global System for Mobile communications [GSM] CAMEL [Customized Application for Mobile network Enhanced Logic]) services. It also plans to integrate mobile phone location services and supports a wide range of man–machine interfaces, such as voice recognition, icons and softkeys. MExE shares several similarities with WAP, in that both protocols have been designed to work with a range of GSM mobile network services from SMS to GPRS and later with UMTS (Universal Mobile Telecommunications System). Whereas WAP incorporates some scripting, graphics, animation and text, MExE allows full application programming, necessitating MExE to include a strict security framework to prevent unauthorized remote access to the user data.

Because the programming and running of Java applications requires the mobile client to have significant processing resources, MExE is primarily aimed at the next generation of powerful smart phones. However, MExE terminals can also include today's regular phones, because MExE incorporates a capability indication method called classmarks, which define the MExE-related services that a particular terminal supports: there will be classmarks that match and those that exceed WAP functionality. The MExE mobile client can inform the MExE server of its classmark and therefore its capabilities.

Development of the initial MExE protocols is being carried out in SMG4 by ETSI, the GSM standards-setting body (see **www.etsi.org**). Supporters of this work include Motorola, Nokia, Lucent Technologies and Nortel (using the Nortel Orbitor smartphone in line with the MExE concept). MExE will considerably outlast WAP because the processing power to run the Java applications is not currently available in mobile terminals. Considering the wide and diverse range of current and future technology and devices that (will) use wireless communication and provide services based thereon, a one-size-fits-all approach is unrealistic. So, in this section we categorize devices by giving them different MExE classmarks:

- MExE classmark 1 – based on WAP – requires limited input and output facilities (e.g., as simple as a 3 line by 15 character display and a numeric keypad) on the client side, and is designed to provide quick and cheap information access even over narrow and slow data connections.

- MExE classmark 2 – based on Personal Java – provides and utilizes a runtime system requiring more processing, storage, display and network resources, but supports more powerful applications and more flexible MMIs.

- MExE classmark 3 – based on the Java 2 Micro Edition Connected Limited Device Configuration (J2ME CLDC) and Mobile Information Device Protocol (MIDP) environment – supports Java applications running on resource-constrained devices.

- MExE classmark 4 – based on the CLI Compact Profile – supports CLI-based applications running on a broad range of connected devices.

7.4.1 MExE classmark 1 (WAP environment)

Classmark 1 MExE devices are based on WAP, a standard protocol to present and deliver wireless information and telephony services on mobile phones as well as other wireless terminals. Supporting the mandatory features of WAP, WAP-enabled devices provide access to the World Wide Web (WWW)-based content for small mobile devices.

7.4.2 MExE classmark 2 (Personal Java environment)

Classmark 2 specifies Personal Java-enabled devices with the addition of the JavaPhone Application Programming Interface (API). The Personal Java application environment is the standard Java environment optimized for consumer electronic devices designed to support WWW content including Java applets. The Personal Java API is a feature level subset of J2SE with some Java packages optional and some API modifications necessary for the needs of small portable devices (e.g., the optimized version of the Abstract Windowing Toolkit targeted at small displays). JavaPhone is a vertical extension to the Personal Java platform that defines APIs for telephony control, messaging, address book and calendar information, etc.

7.4.3 MExE classmark 3 (J2ME CLDC environment)

Classmark 3 MExE devices are based on CLDC with the MIDP. J2ME is a version of the Java 2 platform targeted at consumer electronics and embedded devices. CLDC consists of a virtual machine and a set of APIs suitable for providing tailored runtime environments. The J2ME CLDC is targeted at resource-constrained connected devices (e.g., memory size, processor speed, etc.).

7.4.4 MExE classmark 4 (CLI Compact Profile environment)

Classmark 4 specifies CLI Compact Profile-enabled devices. The CLI environment is a programming language-neutral, Operating System (OS) and CPU-portable environment. It can support applications and services written in a wide range of programming languages (e.g., Visual Basic, ECMAScript and C#). The CLI Compact Profile specifies a minimal set of class libraries, supporting common runtime library features as well as a Web service infrastructure, including Hyper Text Transfer Protocol (HTTP), TCP/IP (Internet Protocol), Extensible Mark-up Language (XML) and Single Object Access Protocol (SOAP). Such devices may not only have limited memory and CPU capability but also limited (or no) display.

The following architectural model shows an example of how standardized transport mechanisms are used to transfer MExE services between the MExE device and the MExE service environment, or to support the interaction between two MExE devices executing an MExE service. Hyper Text Transfer Protocol (HTTP), TCP/IP (Internet Protocol), Extensible Mark-up Language (XML) and Single Object Access Protocol (SOAP). The MExE service environment could, as shown in Figure 7.2, consist of several service nodes each providing MExE services that can be transferred to the MExE device using mechanisms such as (but not limited to) fixed/ mobile/cordless network protocols, Bluetooth, infrared, serial links, wireless-optimized protocols and standard Internet protocols. These service nodes may exist in the circuit-switched domain, packet-switched domain, IP multimedia core network sub-

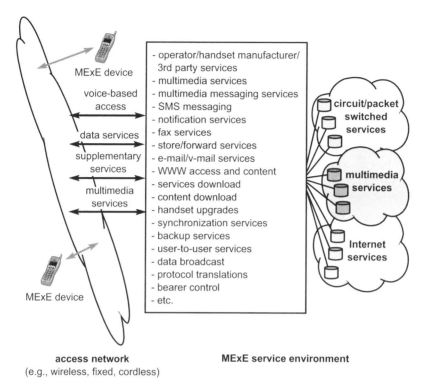

- operator/handset manufacturer/
  3rd party services
- multimedia services
- multimedia messaging services
- SMS messaging
- notification services
- fax services
- store/forward services
- e-mail/v-mail services
- WWW access and content
- services download
- content download
- handset upgrades
- synchronization services
- backup services
- user-to-user services
- data broadcast
- protocol translations
- bearer control
- etc.

MExE device

voice-based
access

data services

supplementary
services

multimedia
services

MExE device

circuit/packet
switched
services

multimedia
services

Internet
services

access network
(e.g., wireless, fixed, cordless)

MExE service environment

Figure 7.2 *Generic Mobile Station Application Execution Environment (MExE) architecure*
Source: 3GPP (2002n)

system or in Internet space (e.g., SMS service centres, multimedia messaging servers, Internet servers, etc.). The MExE service environment may also include a proxy server to translate content defined in standard Internet protocols into their wireless-optimized derivatives. For the versatile support of MExE services, the wireless network must provide the MExE device with access to a range of bearer services on the radio interface to support application control and transfer from the MExE service environment. As MExE also applies to fixed and cordless environments, the MExE device may also access MExE services via nonwireless access mechanisms.

7.5   JAVA 2 MICRO EDITION

Java is known primarily as a server-side programming environment, centred around the technologies that make up the Java 2

Enterprise Edition (J2EE), such as Enterprise Java Beans (EJBs), servlets, and Java Server Pages (JSPs). Early adopters of Java, however, will recall that it was originally promoted as a client-side application environment. In fact, Java was originally designed as a programming language for consumer appliances. This section will outline the key functions of J2ME. In J2ME the Java runtime environment is adapted for constrained devices – devices that have limitations on what they can do when compared with standard desktop or server computers. For low-end devices, the constraints are fairly obvious: extremely limited memory, small screen sizes, alternative input methods and slow processors. High-end devices have few, if any, of these constraints, but they can still benefit from the optimized environments and new programming interfaces that J2ME defines.

Learning about J2ME is not hard: once you understand the new terminology, it is mostly about learning new APIs and learning how to work in constrained environments. (If you think writing an applet is challenging, wait until you try to fit an application into the 30KB of memory some cellphones provide!) With a Java virtual machine, you can use most of the same tools you already use in your code development, and with careful coding you can develop libraries of classes that are portable to any device or computer.

7.5.1 Personal Java and Embedded Java

J2ME is not the first attempt at adapting Java to constrained environments: two other technologies, Personal Java and Embedded Java, made it possible to run Java 1.1.x applications on high-end devices. Personal Java uses the basic Java 1.1 runtime classes and throws in a few features from Java 2. Support for some of the runtime classes is optional, but a Personal Java implementation still requires a couple of megabytes of memory and a fast processor to run; so, it is not a practical solution for truly constrained devices like cellphones and many PDAs. Embedded Java makes every behaviour of both Java VM and runtime classes optional – the implementer can choose exactly which classes and methods are required. There is one limitation, however: the Java runtime environment can only be used by the implementer and cannot be exposed to third parties. In other words, you can use it to write

Java code that runs inside a device, usually as part of the software to control the device, but no one else can write applications for the device. This is done to preserve the "write once, run anywhere" nature of Java, since an Embedded Java environment can do away with fundamental things like runtime class verification and change the public interfaces of core classes. Embedded Java is really only a way to build a "private" Java runtime environment.

Both Personal Java and Embedded Java are being phased out. There is, however, a migration path from Personal Java to J2ME. Although the current version of Personal Java continues to be supported, this is not the case for Embedded Java because J2ME defines suitable small-footprint runtime environments (Ortiz and Giguere, 2001).

7.6 DEVICE INTEROPERABILITY

There are many players in the transcoding business, including Mobixell, Newstakes, Picture IQ and SimplyLook. Consolidation is very likely to happen, as MMS suppliers are more likely to group around a few specialist companies. Interoperability of video services will provide a significant hurdle as manufacturers jostle for the dominant position. Leading phone manufacturers have established relationships to provide client video players on Nokia (Real) and Ericsson (PacketVideo). Although it is still possible to download an alternative player, this will be a considerable barrier for many users.

7.7 DIGITAL RIGHTS MANAGEMENT

There is a significant amount of work being done within the industry from such players as Nokia, who are moving specifications forward in the Open Mobile Alliance (OMA) to address the issues presented by Digital Rights Management (DRM). As we saw in previous chapters, current MMS specifications only allow a proprietary implementation of a forward lock mechanism. OMA's publication of a version 1.0 specification proposes ways in which content can be secured by using WAP. This would support these distribution mechanisms:

- forward lock (or combined delivery of voucher and content);

- separate delivery (of voucher and content);

- superdistribution (allowing forwarding of content to another user).

The specification also formalizes the Mobile Rights Voucher (MRV) object syntax that should be used to define the permissions and rights a user would have over the content downloaded. Nokia have released version 4 of their Mobile Internet Toolkit (NMIT) and have included support for DRM and content publishing.

7.7.1 Differentiate content and control levels

The syntax for the MRV needs to be flexible to accommodate different kinds of content. High-value content needs to have a strict limitation on its reproduction, while no copyright products should be free of a DRM solution. Figure 7.3 shows the relationship between DRM rules and four types of content:

- Premium content – this requires a high level of protection in order to avoid copying. Premium content could include the recent images of a sports event or a ringtone.

- Basic content – this is less expensive and includes horoscopes, comics and premade cards. The DRM solution should be more flexible than for premium content to encourage sharing and traffic. Basic services should offer the exchange of messages

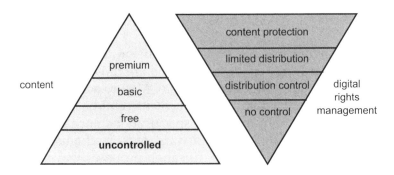

Figure 7.3 Digital Rights Management

among users within certain limits (such as time or maximum number of exchanges).

- Free content – this includes viral marketing and promotional information, such as movie reviews and music samples. This kind of content might still need control (e.g., content providers may want to know who is accessing the content).

- Self-produced content (uncontrolled) – this is one type of content that is unlikely to need a DRM solution, although it may be desirable in the future to allow users to charge for the distribution of their own content, in order to encourage its production.

7.8 M-COMMERCE

MMS will enhance existing m-commerce systems: where it is currently possible to make financial transactions using a mobile phone, MMS will permit interactive experiences where items and services to be purchased can be viewed and examined (Sadeh, 2002). Music could be purchased, with the option of listening to it first. Hotels could be booked and paid for after the subscriber is satisified that it meets his or her requirements. A mobile handset may become the device of choice for downloading content that has been paid for, rather than Internet devices.

7.8.1 Mobile payment options

Mobile payments can be split into three categories: mobile content, out-of-band and proximity. Because of their expertise in the area of billing, network operators are best suited to deliver payment services for mobile content. This type of payment is sometimes referred to as "in-band" because the content and the payment channel are the same (e.g., a chargeable WAP service over GPRS). Applications that could be covered by in-band transactions include video-streaming of sports highlights or video-messaging.

Financial institutions are in the forefront of using mobiles to process "out-of-band" payments – our second category. Out-of-band refers to the fact that the payment channel is separate to that used for a shopping phase. MMS would provide a means for companies that offer a catalogue service (e.g., Argos) to provide an alternative where, instead of using the catalogue, users could browse MMS messages from the store, or remotely. The handset could then be used to make the transaction, allowing the desired product to be delivered or collected. Secure payment methods for the download of music would be welcomed by music companies, along with the necessary DRM systems.

The third potential payment application for mobile commerce is proximity transactions: using the device to pay at a point of sale, vending machine, ticket machine, tolls, parking, etc. With the added functionality of MMS, advertisements and other marketing features could be integrated into this. Operators are generally not interested in providing a standalone payment application for its own sake. Charging and payment are at the heart of their wireless data systems and form part of the infrastructure. The overall experience requires integration at many levels: internal and external content, location services, personalized portal menus and single sign-on systems. Banks are already exploiting the potential of mobile phones to be used as personal secure payment terminals. For example, Orange France has implemented a system whereby a customer uses a Motorola phone with a full size card slot, which accepts the standard chip-based French payment card. The user can conduct a transaction over any channel with a participating merchant and the payment transaction will be sent, via Orange, over the air to the handset as an SMS. At that point a SIM (Subscriber Identity Model) toolkit application is activated and the user inserts his or her payment card, validates his or her PIN locally and a response is returned by SMS to Orange. The user has been securely authenticated and a financial authorization can now take place to check the user has the funds and, dependent on the outcome, the merchant is given a payment assurance.

7.9   WEB SERVICES

A Web service is a software application identified by a Uniform Resource Identifier (URI), whose interfaces and bindings are capable of being defined, described and discovered as XML artefacts. A Web service supports direct interactions with other software agents using XML-based messages exchanged via Internet-based protocols. Mobile Web service interfaces are standardized in various industry fora, such as the OMA, the 3GPP and the WWW Consortium (W3C) of which Nokia is an active member. Mobile Web service interface will be implemented based on widely accepted standard technologies, thus providing a common and interoperable way to define, publish and use mobile Value Added Services (VASs). These technologies are SOAP, XML, WSDL (Web Services Description Language) and HTTP. Initially, the MM7 interface, which provides a method to allow VASs to be integrated with the MMSC (MMS Centre), will be the primary use for Web service technology in the MMS environment, due to its SOAP and XML nature. Mobile Web service interfaces will become an integral part of Nokia's server offering, initially including MMS, terminal management and presence solutions. The MM7 interface became available for the Nokia MMS solution during the first half of 2003.

7.10   3GPP PERSONAL SERVICE ENVIRONMENT (PSE)

Access to MMS services should be independent of access point: Multimedia Messages (MMs) accessible via 3G and 2G mobile networks, fixed networks, the Internet, etc. Common message stores will be an important enabling technology. To facilitate interoperability and universal messaging access, MMS will comply with Virtual Home Environment (VHE): put simply, this lets customers have seamless access with a common look and feel to their services from any environment – home, office, on the move or in any location – just as if they were at home. VHE permits users to manage their services (including non-real-time multimessaging handling) via a user profile, which presents all the different types of media to the user in a unified and consistent manner. MMS supports rich multimedia, and it is therefore important that the concept of a user profile has been included. This user profile is stored in the mobile network and is user-defined and managed via the Internet, determining which MMs are downloaded immediately to the user and which are left on the server for later collection. The user may also choose to receive notifications of certain MM types. The main requirements of VHE are to provide a user with a personal service environment (Penttila, 2003) that consists of:

• personalized services;

- personalized user interface (within the capabilities of terminals);
- a consistent set of services from the user's perspective, irrespective of access (e.g., fixed, mobile, wireless, etc.);
- global service availability when roaming.

VHE is intended to provide:

- a common access for services in future networks;
- an environment for the creation of services;
- the ability to recover a personal service environment (e.g., in the case of loss/damage of user eqipment).

7.10.1 3GPP Generic User Profile

The 3GPP Generic User Profile is the collection of data that is stored and managed by different entities (such as the User Equipment [UE], the Home Environment, the Visited Network and the VAS Provider [VASP], which affects the way in which an individual user experiences services. It is composed of a number of User Profile Components. An individual service may make use of a number of User Profile Components (subset) from the Generic User Profile. The fact of having several domains within the 3GPP mobile system (i.e., circuit-switched, packet-switched, IP multimedia sub-system and the service/application) introduces a wide distribution of data associated with the user. Individual standards already detail some parts of the Generic User Profile. The definition of the 3GPP User Profile is still in its early stages, and will evolve during further releases of the standards.

7.10.2 Liberty Alliance

The Liberty Alliance project is a Sun-led alliance comprising over 60 member companies, formed to deliver and support a federated network identity solution for the Internet (version 1.0 specifications were released in July 2002). This allows people's online identity – their personal profiles, personalized online configurations, buying habits and history and shopping preferences – to be administered by users and securely shared with the organizations of their choosing such as the Liberty 1.0 architecture, which utilizes Web services for communications between providers. Since it is a federated identity solution, different service providers host different parts of users' profiles as appropriate, and users can potentially force data held by one service provider to be shared with other service providers. Identity providers perform authentication and can be chained together, with the active agreement of users, to provide single sign-on.

The Liberty standard is an open technical specification, but users cannot link an identity provider with any service provider or share data between any service providers of their choosing. Providers can only be linked where service and identity providers have chosen to affiliate for mutual business benefit, creating circles of trust. The Liberty architecture does remove the need to rely on a third party to host profiles and the security risk of a single organization hosting entire profiles, such as Microsoft's first attempt at .Net Passport/My Services. However, in its current form it falls short of the ideal network identity infrastructure, since commercial interests look likely to resist the full interlinking necessary for global single sign-on and profile sharing on the Internet and in the mobile MMS world.

7.10.3 Microsoft Passport

In April 2001, Microsoft announced an identity management offering as part of their .Net Web services strategy, launched as .Net Passport/My Services. Users must subscribe to Microsoft Passport, which is a single sign-on system used to access a number of Microsoft services, including Hotmail and the service providers they have signed up. The intention was that service providers could access a potentially broad user profile hosted by Microsoft on their My Services platform, including preferences, eWallet, calendar and alerts. My Services would put users in control on their profiles, by displaying a My Services user interface to enable users to set access rights for different service providers to different parts of their personal data.

After a number of months trying to sign up customers, Microsoft admitted that its business and technical strategy had not been sufficiently successful and withdrew the offering. The problems they encountered included lack of trust over the security of the Passport and hosting valuable customer profile information, and a resistance from organizations to allow Microsoft to host profile data (especially, given its commercial worth). Microsoft also suggested making customers pay for the hosting of profile data, but this is a dubious strategy since users are not accustomed to paying for such services. Microsoft has now gone back to the drawing board, promising that version 2 of Passport will use the trusted Kerberos authentication architecture and that My Services will support federated profile hosting, an approach adopted by the Liberty 1.0 standard (Newbould and Collingridge, 2003).

7.11 IMPROVED BILLING MECHANISMS

The future will bring more advanced billing technques with the view of increasing ARPU. Reverse MMS systems will allow people to be paid mobile credits for viewing MMS messages (specifically, advertising and marketing). With some forms of DRM, a user could create a piece of MMS content and get paid a small credit every time it is passed on by another user. This idea coupled with mobile commerce will enable users to develop content, get "mobile credits" when other users view the content, then use the "mobile credits" to purchase services, such as mobile gaming. Similiarly, users can enter mobile gaming competitions and gain mobile credits for winning the competitions, which again can be used to purchase further services. In this way an entire new economy can be created, based on content and services, which could be extended to allow users to purchase goods using MMS services with the possible consequence of billing systems evolving into banking systems. In the UK, electronic voting has been introduced for local elections. MMS could be used for canvassing and giving details of the nominees prior to voting. The actual vote could be recorded using a centralized system, with the authorities paying for the cost of the messages exchanged, thus necessitating a suitable billing system being in place. Kantsopoulou et al. (2002) contains more details regarding advanced charging and billing systems.

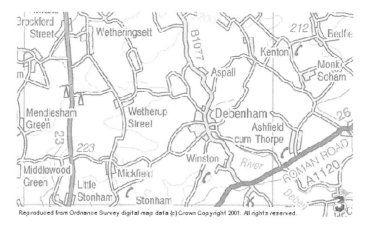

Figure 7.4 Example of a location-based service map
Source: 3

7.12 LOCATION-BASED SERVICES

Another candidate that is eminently suitable for MMS delivery is location-based services. A few operators have already launched location-based services, such as "where is the nearest . . .?". Ironically, some of these services have suffered from a lack of graphical support, such as maps or directions. Location-based services might use locating technologies (network-based, handset-based or hybrid solutions). Alternatively, the service can work if users give details of their own location. With MMS, service operators can deliver maps (see Figure 7.4) and routing instructions. For example, users will be able to locate their closest cinema by looking at a map and receiving advice on traffic and the best route to get there. Services might include:

- maps and routing (how to get to a certain place);

- location of the nearest point of interest (e.g., restaurants, banks, petrol stations or museums);

- information about locations (such as a restaurant review or a tourist guidebook).

It is important for a VASP to be able to deliver location-based services using MMS rather than requiring an application to be downloaded to support the service, which would severely limit the potential market for any service unless the application is pro-vided preinstalled at purchase by the operator.

7.12.1 J-Phone: email me a map

As part of its J-Sky service portfolio, the Japanese operator J-Phone has deployed a commercial wireless concierge service called J-Navi. This directory service includes restaurants, hotels and other business listings, and locates them using colour maps delivered to mobile phones. Launched in May 2000, J-Navi includes more than 17 million businesses, using 256-colour area maps. J-Navi users input the telephone number of the business they want to find and the service delivers a colour map with detailed business listings, including address and Web address. Users can pan, scroll and zoom into the map to find their destination, and then email it as an attachment to themselves or another J-Navi user. The initial goal was to serve 10,000 requests per hour, with a response time on the mobile phone of 2 s or less. J-Navi now serves more than 500,000 customer requests per day, with an average response time of 200 ms. The service is not based on the MMS standard, but users can access the service through the J-Sky wireless portal.

7.13 MULTIMEDIA BROADCAST/MULTICAST SERVICE

The Multimedia Broadcast/Multicast Service (MBMS) introduces the idea of pushing multimedia content to multiple subscribers. Its potential uses are still unknown, but it is anticipated that it could be useful for sporting events, as well as advertising and marketing. SMS currently incorporates this concept in an albeit limited manner, as it is possible to subscribe to such a service as football results and have them sent on to a number of subscribers. The service is still at its specification stage, but a set of guidelines have been published in release 6 of the 3GPP standards (3GPP, 2002h), to aid in the development of the service. The key items are:

• The MBMS architecture will enable the efficient usage of radio network and core network resources, with the main focus on

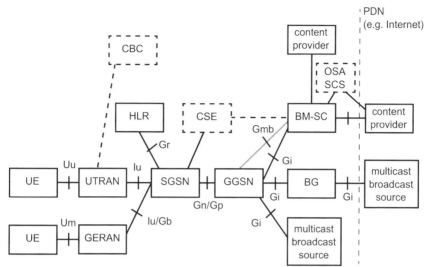

Figure 7.5 MBMS architecture overview

MBMS = Multimedia Broadcast/Multicast Service; PDN = Packet Data Network; CBC = Cell Broadcast Centre; HLR = Home Location Register; CSE = Circuit Switched Environment; BM-SC = Broadcast Multimedia-Switching Centre; OSA = Open Services Access; SCS = Service Capability Servers; UE = User Equipment; UTRAN = Universal Terrestrial Radio Access Network; SGSN = Serving GPRS Support Node; GPRS = General Packet Radio Service; GGSN = Gateway GPRS Support Node; BG = Border Gateway; GERAN = GSM EDGE Radio Access Network; GSM = Global System for Mobile; EDGE = Enhanced Data rate for GSM Extension

radio interface efficiency. Specifically, multiple users should be able to share common resources when receiving identical traffic.

- The MBMS architecture will reuse, to the maximum extent possible, existing 3GPP network components and protocol elements, thus minimizing changes to the existing infrastructure and providing a solution based on well-known concepts.

- MBMS will be a point-to-multipoint bearer service for Internet Protocol (IP) packets in the PS domain.

- MBMS will be interoperable with IETF IP multicast.

- MBMS will support IETF IP multicast addressing.

- MBMS will support different quality-of-service levels. The mechanisms for this need further study (e.g., repetitions to all users).

- Backward compatibility of the MBMS service to the R99 IP multicast delivery mechanism will be considered.

- Interworking possibilities between MBMS-capable network elements and non-MBMS-capable network elements.

In the followong subsections we give details of the components required in an MBMS system.

7.13.1 Broadcast Multimedia-Switching Centre

The Broadcast Multimedia-Switching Centre (BM-SC) is an MBMS data source that can be used for transmission to the user every hour. It offers interfaces over which the content provider can request data delivery to users (the BM-SC will authorize and charge content provider). The Gmb reference point between BM-SC and GGSN (Gateway GPRS Support Node) enables the BM-SC to exchange MBMS service control information with the GGSN (the specification of the Gmb reference point is FFS [For Further Study]). The Gmb reference point exists to carry MBMS service information, but it will not always be necessary to use the Gmb for each service. The possible exceptions are:

- no real-time interface between GGSN and BM-SC – service information is configured or downloaded to GGSN by some other means;

- real-time interface – no per-user authorization (BM-SC pushes or GGSN pulls service information from BM-SC when the service is first activated, but subsequent activations from other users do not require reference to BM-SC);

- real-time interface – per-user authorization (GGSN pulls service information from BM-SC for every service activation);

- no interface between GGSN and BM-SC (service information is configured at GGSN).

7.13.2 MBMS data sources

The architecture allows for other MBMS data sources. Internal data sources may directly provide their data. Data delivery by external sources is controlled by Border Gateways (BG) that may allow, for example, data from single addresses and ports to pass into the PLMN (Public Land Mobile Network) for delivery by an MBMS service. The architecture assumes the use of IP multicast at the reference point Gi. The MBMS data source has only one connection to the IP backbone. The reference point from the content provider to the BM-SC is not standardized and might become too complex or too restrictive for service creation. For example, this may be a reference point between the BM-SC and an authorizing system (the authorizing functionality may be distributed between both entities). The same architecture provides MBMS broadcast services mainly by using transport functions, when user individual SGSN functions are not required; instead, each individual broadcast service is configured in the SGSN.

Once the basic MBMS mechanism is in place, the service possibilities are endless. Coupled with location information, the service is a powerful means of delivering a single piece of content to numerous subscribers efficiently: Maybe team information for a particular football club covering a whole country, or marketing for stores in a particular geographical area. A very specialized service would include broadcasting telemetry and race information at motor racing events. Pictures could be broadcast from points on the circuit that are not visible from the spectator's viewpoint, along with live interviews with experts and commentators. With the introduction of new motor racing rules at the beginning of the 2003 season, information such as fuel load may be of great interest to enthusiasts, although teams may resist this kind of information being available for tactical reasons. Using the dynamic cell size aspects of 3G, racing circuits could own their own cell for the purpose of broadcasting the information and act as a VASP – a means of bringing in more revenue for the events. The service would allow anybody or any organization to set up an MBMS and act as a VASP. Local traffic and weather information could be broadcast. Further services might include financial and business updates, and corporate-wide broadcasts to every employee in a company with information tailored to their own particular location. Religious services could be offered, giving highlights from a church service or prayers from a mosque. Education could feature heavily in future broadcast services by offering exam revision services. These services would allow tutors to actually teach and give demonstrations, rather than just using text, making the concept of learning more appealing.

7.14   CONCLUDING REMARKS

The MMS market is still evolving and it will be some years before its full potential is realized. Until then, the standards will continue to evolve, as will radical services. MMS will make its way into everyday life and will become the normal method of messaging between people and organizations. Although there are competing technologies, the one thing that will ensure its success is the fact that billing and charging mechanisms are already in place, thus encouraging content providers because they will be able to recover costs and make a profit. The idea of paying for content will also be established. This is critical because most competing technologies, most notably those supplied through the Internet, do not have billing systems associated with them. Despite the idea that content is free over the Internet being firmly ingrained in popular culture, subscribers do not appear to have a problem paying for SMS messages.

   Mobile networks provide a convenient method of delivering content that is compact and instantly available (no boot times on mobile phones, as they are always switched on). The manufacturers with expertise in supplying hand-held devices are associated with the mobile industry and thus are going to support the MMS standard, rather than competing standards. For a company such as Nokia which supplies handsets, it is in its interest to support the MMS standards as it also has been instrumental in supplying the infrastructure. People will always have a preference for hand-held devices, because of their practicality, over desktop PCs, as already proven in the messaging space. It costs very little to send an email, yet users are still prepared to pay to send an SMS, due to its convenience and the fact the recipient is more likely to read it straight away. This also applies to MMS: most people want the receiver to see the message as soon as it sent and probably follow it up with a phone call. Broadcast systems will bring about further extensions to the services, allowing content to be pushed to as wide a range of users as possible.

Table of Infrastructure, Content and Software Vendors

12snap (**www.12snap.com**)
12snap is Europe's leading mobile marketing and media sales company, providing campaigns to leading brands and media companies.

3G Lab (**www.3glab.com/products/trigenix.html**)
Trigenix is a development environment for customizable user interface supported on the Symbian 60 OS.

Add2Phone (**www.add2phone.com**)
Add2Phone's mobile advertising server gives you the power to create permission-based marketing campaigns rapidly and cost-effectively in the right mobile media: MMS, Enhanced Messaging Service (EMS) and SMS. All you need to supply are your ideas of how mobility can enhance your marketing strategy.

AgileMotion (**www.agilemotion.com**)
AgileMotion is offering an integrated MMS client, utilizing WAP 1.2.1 or HTTP (TCP [Transmission Control Protocol]/IP) as the data bearer for MMS services. Since MMS is still at the early stage of adoption, it has integrated SMS and EMS format support with our MMS browser (provided as optional modules).

Multimedia Messaging Service   Daniel Ralph and Paul Graham
© 2004 John Wiley & Sons, Ltd   ISBN: 0-470-86116-9

Am-Beo (**www.am-beo.com**)

Am-Beo is a global provider of key software solutions for the telecommunications and financial services markets. It creates software specifically for the intricate and complex issues associated with pricing, rating and invoicing of next generation, content products and services. Included in Am-Beo's offerings is the capability to manage the digital supply chain and to provide a means to manage the settlements required between the content providers and the network operators. Am-Beo is a product-focused company, delivering software that is designed for integration and rapid deployment, and to work within an existing infrastructure.

ANAM (**www.anam.com**)

ANAM delivers an Enterprise Wireless Internet product that is specifically designed for Windows NT and 2000 server platforms. The WirelessWindow product includes a combined WAP 1.1 and SMS gateway and boasts an excellent price/performance metric.

Akumiitti (**www.akumiitti.fi**)

Allato Technologies (**www.alatto.com**)

Memphis is a suite of products designed to help maximize the revenue potential of MMS. Together these provide a full suite of functionality to create a complete user experience for MMS messaging.The Memphis Suite supports flexible charging events and each is highly configurable. The Memphis products have been designed with reliability, security and scalability in mind and have been validated with MMSCs from the major vendors.

ArcSoft (**www.arcsoft.com**)

ArcSoft's mobile and wireless solutions deliver a world-class platform for developing and deploying compelling, differentiated multimedia services to exploit this market opportunity.

Argogroup (**www.argogroup.com**)

Do you know if your customer's MMS will be delivered and displayed correctly on all MMS-enabled devices? Check with UbiquinoX Monitor Master – a software package that can be used by network operators, content and application developers and Quality Assurance (QA) professionals to test and monitor the quality and usability of wireless data services, as viewed by Wireless Mark-up Language (WML), SMS, EMS and MMS-enabled devices.

BTexact (**www.f-nets.com**)
A mobile video streaming player and content server for Symbian and PPC (Pocket PC) formats.

bmd wireless (**www.bmdwireless.com**)
The Wireless Application Messaging Server (WAMS) is the most widely deployed cross network SMS platform that enables any mobile user from any wireless network in the world to reach an application or service.

BitFlash (**www.bitflash.com/**)
BitFlash offers standards-based software for wireless multimedia messaging. Its mobile Scalable Vector Graphics (SVG)-based solutions are used by device manufacturers, content creators, wireless carriers, infrastructure solution vendors and handset manufacturers to create, deliver and predictably render visually rich information and entertainment content.

Brainstorm (**www.brainstorm.co.uk**)
Brainstorm is Europe's leading specialist in developing and managing value-added mobile services. It works primarily with marketing agencies, portals and network operators, providing a one-stop solution for companies wanting to utilize mobile technologies to raise brand awareness, build customer loyalty and generate revenues.

Comverse (**www.comverse.com**)
Total communication creates a borderless environment where people are free to communicate in the way that is most appropriate and convenient for them. Talking, voice messaging, emailing, text messaging, multimedia messaging, chatting and conferencing all become equally accessible options.

CinemaElectric (**www.cinemaelectric.com**)
CinemaElectric is the producer and distributor of a unique form of Downloadable Digital Cinema called PocketCinema™ to be played on mobile phones, handheld computers, desktop computers and future movie-playing personal electronics.

Clickatell (**www.clickatell.com**)
Clickatell has brought itself to the forefront of global mobile messaging by consistently ensuring the broadest global network coverage, unmatched delivery reliability and service levels. In addition to providing a robust and non-complex platform for integration

with any messaging application, Clickatell has created a suite of communications–marketing products that facilitates targeted messaging to recipients using mobile devices. There is easy connection to Clickatell's global SMS gateway through HTTP, SMTP, File Transfer Protocol (FTP), Extensible Mark-up Language (XML) and Com Object, all of which can be done in real time.

Critical Path (**www.cp.net**)
Critical Path works with strategic partners to deliver complete MMS solutions. It provides key technology components required for all MMSs, including a highly scalable messaging server, centralized address book and multimedia file storage and sharing.

Cybird (**www.cybird.co.jp/english**)
Mobile content business development and operation of pay content for Internet-enabled mobile phone/Personal Handyphone System (PHS) official portal sites and development of mobile content commissioned by third parties.

Deltica (**www.delicta.com**)
Deltica's SMS gateway service allows you to send SMS messages (supports OTAP [Over The Air Protocol] to Nokia 7110) from your webserver or email software to mobile telephones anywhere in the world, with per-message pricing and no set-up costs.

Dialogue Communications (**www.dialogue.co.uk**)
Content delivery over MMS, SMS and WAP, wireless email (SMS, WAP and GPRS) and multimedia messaging.

Digital Mobility (**www.digimob.com**)
Wireless Cube "plug and play" wireless SMS messaging and WAP gateway solution. The Wireless Cube supports a powerful family of SMS messaging and WAP services, enabling organizations to roll out wireless messaging extensions to their own corporate networks quickly, efficiently and economically.

eMood (**www.emoodsoft.com**)
eMood Messenger is the first EMS/SMS messenger for Palm OS. It sends and receives EMS/SMS messages (formatted text messages with pictures, animations and melodies). It supports EMS over Bluetooth on your Palm today (and it works!). It also supports any interface available on a Palm device (serial connection Infrared or Bluetooth). It exchanges messages with GSM EMS/SMS mobile

phones and other palm devices and communicates with most GSM phones, such as Ericsson, Nokia, Siemens and Motorola.

Empower Interactive Group (**www.eigroup.com**)
This messaging gateway is a high-capacity corporate-to-carrier grade IT solution for complete MMS/SMS messaging functionality. It has an MMS/MM7 client interface.

Ericsson (**www.ericsson.com/mms/**)
Ericsson's MMS is an end-to-end solution enabling operators to deliver a rich MMS experience to customers.

Exago (**www.exago.com**)
Exago's products combine the latest knowledge management expertise with practical collaboration, advanced search and summarization tools to deliver value to your entire organization. Exago delivers scalable, practical, forward-thinking, user-friendly solutions that deliver immediate value.

Eyematic (**www.eyematic.com**)
Eyematic's advanced multimedia software platform powers the new generation of visual applications and services across mobile and fixed networks. Its platform is used by mobile network operators, mobile handset manufacturers, content publishers and the entertainment industry to create a wide range of communications, entertainment and information products and services.

Exomi (**www.exomi.com**)
Exomi provides carrier-grade wireless gateway solutions for WAP, SMS and wireless email access: SMS gateway, wireless mail solution, WAP gateway, WAP Push Proxy Gateway (PPG).

Followap (**www.followap.com**)
Followap offers advanced solutions for mobile operators that are at the forefront of existing and emerging second-and-a-half generation (2.5G) and third-generation (3G) technologies. Followap brings value to customers with next generation messaging and presence services, covering wide subscriber segments in the areas of presence, instant messaging, chat, location, multimedia messaging, voice and mobile entertainment.

Funmail (**www.funmail.com**)
Message Router seamlessly connects diverse applications to different types of messaging (MMS, SMS, EMS) centres.

First Hop (**www.firsthop.com**)
The messaging manager manages an evolving service portfolio in a multivendor, multistandard messaging environment.

Freever (**www.freever.org**)
Freever is a new media: a means of communication and the link chosen by the mobile generation. Today, there are already hundreds of thousands of users of mobile phones who have chosen Freever for their daily communication and information needs. Who are they? Typically, the 15–25-year-olds who found in Freever a new media.

Futurice (**www.futurice.com**)
Futurice is a leading provider of advanced communication solutions for the mobile environment and one of the first companies in the world to concentrate solely in visual mobile messaging.

Helix (**www.helixcommunity.org**)
The Helix community is a collaborative effort between Real-Networks and independent developers and offers a digital media delivery platform including client, producer and server.

Intec (**www.intec-telecom-systems.com**)
InterconnecT is the world's most successful carrier-to-carrier billing system, offering the capability to rate and bill for all kinds of traffic and network events, including IP and 3G needs. InterconnecT is supported by a range of additional modules, such as auto-reconciliation and links to ERP (Enterprise Resource Planning) systems that make it easy to implement in any environment. Other related product offerings include InterconnecT ITU (International Telecommunications Union) for settlement based on ITU guidelines and the InterconnecT ASP (Application Service Provider) service.

Intellisoftware (**www.intellisoftware.co.uk**)
SMS gateway software for use with a two-way SMS gateway, SMS-to-email, email-to-SMS and COM (Common Object Model) component for ASP, VB (Visual Basic) and C++, utilizing GSM modems, handsets or TAP (Telecator Alphanumeric Protocol) protocol.

Internexium (**www.internexium.com**)
The VisibleMessage suite enables you to execute extremely fast and easy creation of MMS. In one easy, understandable suite the

complex mobile network technology is made completely transparent for the user.

iPlanet (Sun-Netscape) (**www.iplanet.com**)
iPlanet's portal server (Mobile Access Pack 3.0) makes portal content available to mobile users from any computing device. The Mobile Access Pack enables users to access the full array of personalized content and services provided by the iPlanet portal server, which might include: email, calendar, address book, news, stock quotes, weather and other content, location-based services, SMS, enterprise information and applications.

iScan (**www.iscan.it**)
iScan specializes in data capture and acquisition software for Ericsson handsets. iScan's application designer software facilitates the retrieval of menu applications from a remote server via SMS. The device supports various mobile-scanning applications.

Kannel (**www.kannel.org**)
Open source WAP/SMS gateway.

Kuulalaakeri SMS-gateway (**www.kuulalaakeri.com**)
Kuulalaakeri's QLA messaging server is a powerful application platform for operators and service providers. It also supports Nokia CIMD (Computer Interface to Message Distribution), CMG EMI, Logica SMPP (Short Message Peer to Peer) and Sema SMS2000 protocols. Easy API (HTTP) offers a fast way to develop one- and two-way SMS with text, binary (logos and ringing tones), unicode and EMS content.

Jataayu software (**www.jataayusoft.com**)
WAP gateway for MMS clients.

Java SMS library project (**javasmslib.sourceforge.net**)
A multipurpose library that sends short messages, images, ringtones and OTA provisioning over TCP/IP, modem or direct cable connection to a GSM phone. The Java SMS library is a donation to the open source code community of new-phone.com.

LiquidAirLab (**www.liquidairlab.com**)
LiquidAirLab is a world-leading developer of carrier-grade solutions for visual MMS messaging on wireless networks. Its MMS studio contains all the applications that are needed for profitable MMS sevices: dynamic MMS composer, multimedia album, trans-

coding engine, transaction engine and XML-based content integration engine.

LogicaCMG (**www.logicacmg.com**)
LogicaCMG provides world-class fixed and mobile operators, whether established or new entrants, with end-to-end consulting, program management, systems integration and operational support, together with a comprehensive portfolio of industry-leading products and solutions. These include messaging, prepaid, radio planning, billing, mobility and payments, as well as OSS and CRM solutions.

LockStream (**www.lockstream.com**)
LockStream is a leading provider of software for the secure distribution of media. Its proprietary technologies provide the highest levels of security in the industry, offer unparalleled ease of use and are extremely customizable. Additionally, because of LockStream's focus on creating a compact, efficient and multiplatform solution, its DRM software operates on the widest range of platforms, including mobile phones, PDAs, set-top boxes, automobile entertainment systems, PCs and other media access devices.

LightSurf (**www.lightsurf.com**)
LightSurf is the Instant Imaging Company whose core technology, the LightSurf IV Instant Imaging Platform, is a suite of carrier-grade MMS services that allows users to capture, view, annotate and share MMs with any handset or email address, regardless of device, file type or network operator. LightSurf IV also offers premium content delivery services that provide users with news, finance, sports and other time-sensitive visual updates, including syndicated third-party image and audio content. LightSurf's Instant Imaging Platform has been deployed globally with customers such as Sprint PCS, Kodak, O_2, KDDI, Motorola and many others.

mBlox (**www.mblox.com**)
mBlox is Europe's leading provider of SMS infrastructure services to the content and applications industry. It makes the task of sending and receiving SMS simple, reliable and cost-effective, freeing its clients from the commercial and technical complexities that can cost so much time and effort. Companies use its standard and premium-rate outbound and inbound services to provide information, entertainment and communications services to consumers or enterprise staff.

Materna (**www.annyway.de**)
Anny Way's MMS value-added service applications introduce a new generation of mobile communication. Now users of mobile phones/smartphones can combine colour pictures and animations, sounds and text in one message. With MMS, mobile messaging becomes more personal and versatile.

mmO₂ (**www.sourceO2.com**)
Source O_2 has been designed to help you get your product to market with O_2 as efficiently and effectively as possible. On the site you will find information about its technologies and networks, information on the conferences it holds every four months, developer news and, most importantly, how to work with mmO_2 to reach its 17.75 million customers.

MobileWay (**www.mobileway.com**)
MobileWay is heavily involved in the development of MMS, particularly in the billing and payment aspect of the next generation messaging standard. One of the main challenges facing operators, handset manufacturers and application developers for MMS alike has been how to bill for multimedia content. MobileWay is leveraging its legacy of privileged relationships with the world's mobile operators and global payment brands to work on a number of solutions. It is currently using premium rate SMS and existing billing relationships to build the most effective and economically viable solution for the wireless industry.

MindMatics (**www.mindmatics.co.uk**)
MindMatics offers a unique portfolio of interactive advertising solutions based on permitted marketing and loyalty programs. MindMatics' concept is that users actively sign up for sponsored services with their personal profile and are always in control of what type, when and how much advertising they are exposed to.

mi4e (**www.mi4e.com**)
mi4e is one of the leading providers of messaging and device management software and services to the world's largest carriers, virtual operators and content service providers. mi4e's Embedded Mobilization technology enables wireless carriers to develop and deliver secure, real-time, two-way messaging solutions on any mobile device across multiple network protocols.

NewBay (www.newbay.com)
NewBay Software is the leading provider of phone blogging (i.e., using a website on a daily basis to record personal events) products and services that enable mobile operators to increase revenues, customer loyalty and Internet data services. FoneBlog™ is the most exciting and fun way for mobile phone users to create and maintain their own personal website.

NetMMS (de) (http://demo.net-communications.de/netmms)
German MMS content site and portal provider.

NCL Technologies (www.ncl.ie)
NCL's SwiftMedia is a software development toolkit for wireless application developers that allows you to develop custom media message (EMS/SmartMessage/MMS) applications quickly.

Nextreaming (www.nextreaming.com)
Nextreaming Corporation provides the necessary software applications and components to bring rich media services to mobile environments. One is Nexencoder™ which compresses video. The other, Nexplayer™, is an optimized multimedia player for mobile handsets, PDAs and PCs.

Nethix (www.wapnethix.com)
WEB-Engine is the first embedded HTML (Hyper Text Mark-up Language), WAP and SMS server over the GSM network on the market. WAP-Link is a software library that implements the WAP protocol stack and allows the integration of a single user WAP and SMS server on the target application. Following the huge success of the WAP-Engine, Nethix is releasing a brand new model for industrial applications. Called, GSM-Term, this is a modem specifically designed for industrial applications.

Network365 (www.network365.com)
This is mZone's mobile commerce server for WAP, i-mode and SMS.

Nokia (www.forum.nokia.com/tools)
Nokia's Mobile Internet Toolkit 3.1 is part of an end-to-end solution from Nokia for development and deployment of MMS services by operators and content providers.

Now (www.nowmms.com)
The NowMMS.com service is built on top of the Now SMS/MMS

gateway and Now MMSC and delivers affordable SMS and MMS solutions that can be deployed independently of mobile network operators. Free trial versions of the Now SMS/MMS gateway and built-in MMSC are available.

OctaneMobile (**http://www.octanemobile.com**)
The Octane MMS suite provides network operators with the core competencies needed to collaborate with content providers to deploy dynamic MMS content alerts and services. The product suite facilitates the launch of a network operator's MMS content partner program and overcomes the partner access, service creation, content delivery and message tracking hurdles associated with extending MMS infrastructure to third parties.

Oplayo (**www.oplayo.com**)
Oplayo develops easy-to-use streaming video technology in a single format for enterprises, content holders and service providers.

Oratrix (**www.oratrix.com**)
Oratrix has been helping create the Web's most complete streaming media presentations since 1998. Its GR*i*NS editing technology has been recognized to be the most powerful Synchronized Multimedia Integration Language (SMIL) editor available on the market, and its SMIL 1.0/2.0 players set the standard for language completeness and compliance with the standards specification. This is because it not only follows SMIL standards, it helped create them.

Openwave (**www.openwave.com**)
The Openwave MMSC is a high-capacity, massively scalable service platform that brings the multimedia messaging capabilities of the Internet to mobile devices, while retaining the immediacy and simplicity of the mobile SMSs that are popular in many parts of the world today. It enables handset users to send and receive messages that contain text, music, graphics, video and other media types – all in the same message.

PacketVideo (**www.packetvideo.com**)
PacketVideo is the first company in the world to demonstrate MPEG4 video image streaming to mobile devices. Its standards-compliant, carrier-grade technology is being deployed around the world today and its breakthrough technology platform, called the pv3 Mobilemedia System, provides encoding (author), serving (server) and decoding (player). The pv3 Mobilemedia System is

the powerhouse behind complete solutions for mobile telecommunications service providers, media programming providers and application developers.

Peramon Technology (**www.peramon.com**)
The MOBILIZER is the world's most comprehensive mobile Internet platform and consists of a complete set of mobile data access applications and infrastructure products based on SMS, WAP and VoiceXML technologies. It gives employees secure access to their personal and business applications to feed their working lifestyles irrespective of the mobile device they choose to use.

PictureIQ (**www.pictureiq.com/**)
Those who prepare and deliver content and Internet-enabled applications must take into account an assortment of network-connected devices and access points. Until PictureIQ, no automated solution existed specifically to accommodate the conversion of visually rich content that efficiently and cost-effectively met this demand.

Pixer (**www.electricpocket.com/products/pixer.html**)
Electric Pocket is a leading provider of wireless applications and development services for Palm OS and Pocket PC-based mobile devices. It has developed leading-edge mobile and wireless solutions for companies, including Handspring, Palm, Ericsson and Ubinetics.

PocketThis (**www.pocketthis.com**)
PocketThis delivers software that allows mobile operators to rapidly partner with content and service providers, create new and useful services, and generate increased mobile data services revenue.

Quartez (**www.quartez.com**)
Quartez offers intelligent mobile and business process integration using text messaging. It builds, implements and hosts messaging solutions for intelligent mobile and business process integration.

RealPlayer (**www.realnetworks.com/**)
Ericsson, the world's largest supplier of mobile systems, is to integrate RealNetworks' Helix™ media delivery technology into the new Ericsson Content Delivery System (ECDS). This combination

will enable operators to purchase one comprehensive, end-to-end solution to create mobile media services for consumers.

SenseStream (**www.sensestream.com**)
SenseStream provides critical multimedia management technology that many mobile multimedia services and applications rely on to successfully and seamlessly deliver multimedia messages. Its advanced and powerful multimedia content adaptation server, DynaFit^{TM}, is used by many of Asia's leading wireless operators for deployment of successful mobile multimedia services. Sense-Stream's proprietary PONFA^{TM} (Publish Once N Fit All) server and publishing tools allow content providers to create and publish wireless multimedia contents.

Stalker (**www.stalker.com**)
Stalker Software is a recognized leader in the development of high-end Internet messaging infrastructure. The CommuniGate Pro messaging servers empower enterprises, Internet Service Providers (ISPs) and telcos with sustainable competitive advantage, improved profitability and lower cost of ownership by providing cutting-edge technology. Designed to provide performance, availability and features "required tomorrow", Stalker Software solutions ensure that customers can exchange information where it is needed and when it is needed more efficiently, more effectively and more reliably than with any other solution available.

Simplewire (**www.simplewire.com**)
Simplewire is a wireless messaging infrastructure and software provider. Its unique platform and accompanying tools ease the process of creating wireless applications for businesses, telecommunications carriers and software developers. Its wireless messaging network currently supports over 300 networks in 118 countries. Simplewire's developer program offers developers the opportunity to create and quality-assure their wireless messaging application before commercially releasing it. The program is free of charge and provides wireless development tools and resources to help developers get started. Developers may create an account and then receive a developer subscriber ID, giving the developer access to evaluation versions of Simplewire's SMS software development kit for use with Perl, Java, ActiveX, Shared-Object and PHP.

SmartTrust (**www.smarttrust.com**)
SmartTrust is a leading developer of infrastructure solutions de-

signed to provide mobile subscribers with easy-to-use and secure personalized services. Using SmartTrust's solutions, mobile operators are able to manage and optimize both handsets and GSM/3G (U)SIM cards to deploy revenue-generating Value Added Services (VASs).

SMS Client (**www.styx.demon.co.uk**)
This is an open source client implementation for Cellnet's GSM Short Message Service Centre (SMSC) using TAP.

SMILGen (**www.smilgen.org/index.html**)
SMILGen is a SMIL (and XML) authoring tool designed to ease the process of XML content creation. SMILGen understands XML syntax and handles the nesting and formatting of XML. This allows authors to worry about the content that they are trying to author without having to remember each quote and closing brace. SMILGen also understands the languages it authors and knows what attributes a specific element uses or what child elements a given element may contain. Both of these features help eliminate a number of common XML syntax errors as well as making it easier to edit without having a reference to the language right by your side.

Software Scientific (**www.scientific.co.uk**)
Email to SMS Summarizer is a system that intelligently converts emails (or text) into SMS messages that are no longer than 160 characters and preserves the meaning and intelligibility of the original message. It aggressively (and iteratively) compresses email summaries using SMS short forms to maximize content while preserving clarity. InterAct is a system for interacting with end-users via SMS messages that is knowledge-driven, highly configurable and suitable for driving marketing campaigns. AutoAnswer is a system for providing customer support via SMS, email or the Web that once again is knowledge-driven.

Soprano (**www.soprano.com.au**)
High-performance, reliable platforms for messaging (MMS, EMS, SMS, WAP, IVR [Interactive Voice Response]) rated at over 10,000 messages per second and vertical platform support for content aggregate/deploy, M-commerce, premium services, corporate messaging and mobile games.

Symbian (**www.symbian.com**)
Symbian is a software licensing company that is owned by wireless industry leaders and is the trusted supplier of the advanced, open, standard OS – Symbian – for data-enabled mobile phones.

TOM (jp) (**www.mms.tom.com**)
Japanese MMS content site.

Tornado (**www.mobiletornado.com**)
Mobile Tornado develops integrated voice and data solutions that are accelerating the deployment of data services over current wireless networks and devices.

ThunderWorx (**www.thunderworx.com**)
The ThunderMMS platform offers a complete and highly stable end-to-end multimedia messaging system. Its flexible architecture facilitates its integration with any existing infrastructure and can be customized to have the look and feel that you desire. ThunderMMS Studio incorporates all the necessary tools required to create, preview and send MMs.

Tricomtek (**www.tricomtek.com**)
Developers of short messaging applications for the 3G SMSC and SMS platform that enables subscriber access to both the CDMA and 3G/UMTS (Universal Mobile Telecommunications System) networks. mobilePOST-MMSC is designed for 2.5G and 3G services, enabling wireless carriers and service providers to deliver MMs to customers. Tricomtek's MMSC has a modular basis architecture that enables access to its customized MMS services library (written in C).

VoiceAge (**www.voiceage.com**)
The AMR (Adaptive Multi Rate), a mandatory 3G speech compression codec, is based on VoiceAge ACELP (Algebraic Code Excited Linear Predictive) proprietary technology. Users can download free of charge the xde decoder to play back their PC audio content.

Volubill (**www.volubill.com**)
Billing infrastructure supplier.

Visualtron Software (**www.visualtron.com**)
Visualtron Software builds and markets software infrastructure for deploying and managing wireless Internet, GPRS and SMS applications on mobile devices, such as VisualGSM SMS server.

Watercove (**www.watercove.com**)
WaterCove Networks has introduced the industry's first Mobile Data Service System that overcomes the performance, provisioning, charging and partnership obstacles inherent in operators' routing-centric legacy networks and equipment.

wapMX (**www.wapmx.com**)
wapMX provides a WAP hotmail gateway, a Post Office Protocol version 3 (POP3) email gateway. SMS servers, SMS brokers, OTA provisioning, logos, ringtones, vCards and vCalendar SMS applications as well as WAP and SMS software development.

XIAM (**www.xiam.com**)
The XIAM Information Router allows for the intelligent routing of information between an enterprise and mobile users over the SMS. Remote users can easily update a database, reply to emails, access an intranet or Internet web page or even restart a critical program from any standard mobile device, at any time.

X-Smiles (**www.x-smiles.org**)
X-Smiles is a Java-based XML browser, intended for both desktop use and embedded network devices and to support multimedia services. The main advantage of the X-Smiles browser is that it supports several XML-related specifications and is still suitable for embedded devices supporting the Java environment.

Zidango (**www.zidango.com**)
Zidango is a mobile portal where you can download MMS templates and messages to your mobile phone. An online user album will be created automatically for you when your MMS account is activated at Zidango.

Glossary

2G Second Generation mobile communication systems. Examples include GSM, TDMA, PDC and cdmaOne.

2.5G Second and a Half Generation mobile communication systems. Umbrella term used to designate intermediate packet-switched, always-on systems. Examples include GPRS and cdma2000 1x.

3G Third Generation mobile communication systems. Key features of 3G systems, as originally recommended by the International Telecommunication Union (ITU) as part of its International Mobile telecommunications 2000 initiative (IMT-2000), include a high degree of commonality of design worldwide, worldwide roaming capability, support for a wide range of Internet and multimedia applications and services, and data rates in excess of 144kbps. Examples of 3G standards are WCDMA, UMTS, EDGE and cdma2000.

3GPP/3GPP2 Third Generation Partnership Projects in charge of laying out the specifications of 3G standards. The 3GPP partnership focuses on WCDMA/UMTS and EDGE, while 3GPP2 focuses on cdma2000.

3GSM Third generation services delivered on an evolved core GSM network. 3GSM services are delivered at a technical level on third generation standards developed by 3GPP, which utilize air interfaces for WCDMA and, in some specified markets, EDGE.

ACELP Algebraic Code Excited Linear Prediction. A process employed in analysis by synthesis codecs in order to predict the filter coefficients required to synthesize speech at the receiving party. This process is employed in the ITU-T (International Telecommunications Union-Telecommunication Standardization Sector) G723.1 codec specification.

Multimedia Messaging Service Daniel Ralph and Paul Graham
© 2004 John Wiley & Sons, Ltd ISBN: 0-470-86116-9

AMR   Adaptive Multi Rate. A codec offering a wide range of data rates. The philosophy behind AMR is to lower the codec rate as the interference increases and thus enabling more error correction to be applied. The AMR codec is also used to harmonize the codec standards amongst different cellular systems.

API   Application Programming Interface. A set of calling conventions which define how a service is invoked through a software package.

APN   Access Point Node. An Access Point Name provides routing information for SGSN (Serving GPRS Support Nodes) and GGSN (Gateway GPRS Support Nodes). The APN consists of two parts: the Network ID, which identifies the external service requested by a user of the GPRS service and the Operator ID, which specifies routing information.

Application   A piece of software that performs specific functionality. It can reside on the terminal device or in the network accessed by the device, or it can be distributed between both.

ARPU   Average Revenue Per User. A calculation often used to determine the overall value of an application. It is also used to rate particular customers, especially in the wireless space, by comparing someone's account with the overall average.

ASP   Active Server Pages. Bandwidth. A measure of information-carrying capacity on a communications channel. The greater the bandwidth the greater the information-carrying capacity. Bandwidth is measured in hertz (analogue) or bps (digital).

Base station   A transmitter and receiver in a single piece of hardware (the .xed part of a radio cell) used for communicating with mobile terminals.

BG   Border Gateway. A Border Gateway function terminates the Gp interface to a PLMN (Public Land Mobile Network). This function is typically an edge router supporting the BGP (Border Gateway Protocol) and security protocols such as IPSec (IP Security).

Bluetooth   A short-range radio interface, designed to allow different devices to communicate with each other without needing to be connected by wires. It coexists with WAP and other network technologies and is supported by 1,200 organizations. The standard range of the radio link is 10 m, although there is an optional extension to 100 m.

BM-SC   Broadcast Multimedia-Switching Centre.

Branding   A form of marketing that is designed to make users aware of a company's name, logo or catchphrase. The purpose is to build an image, rather than promote a specific product, offer, promotion or service.

Broadband   A term used to describe a high-speed communication service. It is generally taken to include channels with a bandwidth capacity of more than 144kbps.

CAD   Computer Aided Design. Using computers to design products. CAD systems are high-speed workstations or desktop computers with CAD software.

Calling party pays   Charging regime whereby the charge for a phone call or message is paid by the party that initiates it.

CAMEL   Customized Application for Mobile network Enhanced Logic. CAMEL is a 3GPP (Third Generation Partnership Project) initiative to extend traditional IN (Intelligent Network) services found in fixed networks into mobile networks. The architecture is similar to that of traditional IN, in that the control functions and switching functions are remote. Unlike the fixed IN environment, in mobile networks the subscriber may roam into another PLMN (Public Land Mobile Network), consequently the controlling function must interact with a switching function in a foreign network. CAMEL specifies the agreed information flows that may be passed between these networks.

CAP   CAMEL Application Part. CAMEL Application Part is a real-time protocol used to support the information flows between CAMEL (Customized Application for Mobile network Enhanced Logic) functional elements such as the SCF (Service Control Function) and SSF (Service Switching Function).

CBC   Cell Broadcast Center. The Cell Broadcast Centre is the functional entity within the mobile network that is responsible for the generation of cell broadcast information.

CC/PP   Composite Capabilities/Personal Preferences. The capabilities of devices and preferences of users can be stated in CC/PP and used to select profiles for presenting web information on devices such as mobile phones with limited capability – supports the content/presentation distinction.

CCS7   Common Channel Signalling 7. A CCS (Common Channel Signalling) system defined by the ITU-T (International Telecommunications Union-Telecommunication Standardization Sector). SS7 is used in many modern telecoms networks and provides a suite of protocols that enables circuit and non-circuit-related information to be routed about and between networks. The main protocols include MTP (Message Transfer Part), SCCP (Signalling Connection Control Part) and ISUP (ISDN User Part).

CDMA 1X   An upgrade to CDMA networks that enables them to deliver packet data services and to provide access bandwidth of around 5,060kbps (in theory, speeds up to 144kbps are possible, but 60kbps is the limit of most practical implementations). The full name of this technology is CDMA 1X-RTT.

CDMA   Code Division Multiple Access. A method of frequency reuse whereby many radios use the same frequency, but each one has a unique code. GPS uses CDMA techniques with Gold's codes for their unique cross-correlation properties.

cdma2000   A CDMA version of the IMT-2000 standard.

cdmaOne   The original ITUIS-95 (CDMA) mobile interface protocol that was first standardized in 1993. This is a 2G mobile technology.

CDR   Charging Data Record. A database record unit used to create billing records. A CDR contains details such as the called and calling parties, originating switch, terminating switch, call length, and time of day. When applied to GPRS CDR are generated typically by the SGSN (Serving GPRS Support Node) and GGSN (Gateway GPRS Support Node), recording data volumes and QoS (Quality of Service) rather than call time. These records are passed to the CGF (Charging Gateway Function) for consolidation prior to being passed to the billing platform.

Cellular   Wireless communications based on cells, each one defined by the transmitting and receiving range of a base station.

CGF   Changing Gateway Function. Element with a GPRS network that consolidates, filters and optimizes CDR (Call Detail Record) prior to their transmission to the Billing Platform. This function may be distributed within the SGSN (Serving GPRS Support Nodes) and GGSN (Gateway GPRS Support Node) or centralized.

CGI   Common Gateway Interface. A relatively compact program written in a language such as Perl, Tcl, C or C++ that functions as the glue between HTML pages and other programs on the Web server.

CHTML   Compact HTML. A subset of HTML for cellphones and PDAs. Developed by NTT Docomo for its i-Mode wireless system in Japan, CHTML is designed for the limited screen displays and functionality of hand-held devices.

CIMD   Computer Interface to Message Distribution. CIMD is a dedicated data exchange protocol for connecting application programs to the Nokia Artuse service center, as well as the short message service centre (SMSC) and the USSD centre (USSDC).

CLDC Connected Limited Device Configuration. This is one of two configurations defined through the Java Community Process as part of Java 2 Platform, Micro Edition (J2ME). CLDC is the foundation for the Java runtime environment targeted at small, resource-constrained devices, such as mobile phones, pagers and mainstream personal digital assistants.

CLI Common Language Infrastructure. This provides a virtual execution environment comparable with the one provided by Sun Microsystems for Java programs. In both environments, CLI and Java use a compiler to process language statements (also known as source code) into a preliminary form of executable code called bytecode.

CLP Command Line Protocol.

CN Core Network. This is an evolved GSM Core Network infrastructure or any new UMTS Core Network infrastructure, integrating circuit and packet-switched traffic. This may include such functions as the MSC (Mobile Switching Centre), VLR (visitor Location Register), HLR (Home Location Register), SGSN (Serving GPRS Support Node) and GGSN (Gateway GPRS Support Node).

COM Component Object Model. A component software architecture from Microsoft, which defines a structure for building program routines (objects) that can be called up and executed in a Windows environment. This capability is built into Windows 95/98 and Windows NT 4.0. Parts of Windows itself and Microsoft's own applications are also built as COM objects. COM provides the interfaces between objects, and Distributed COM (DCOM) allows them to run remotely.

CORBA Common Object Request Broking Architecture. A standard from the Object Management Group (OMG) for communicating between distributed objects (objects are self-contained software modules). CORBA provides a way to execute programs (objects) written in different programming languages running on different platforms no matter where they reside in the network. CORBA is suited for three-tier (or more) client/server applications, where processing occurring in one computer requires processing to be performed in another. CORBA is often described as an "object bus" or "software bus", because it is a software-based communications interface through which objects are located and accessed.

CPIM Common Presence and Instant Messaging. An abstract protocol from the IETF that specifies behaviours and information flow of an instant messaging (IM) system. Designed as a common blueprint for building gateways, any CPIM-compliant IM system will work with any other CPIM-compliant system.

CRM   Customer Relationship Management. An integrated information system that is used to plan, schedule and control the presales and postsales activities in an organization.

CSCF   Call Server Control Function.

CSD   Circuit Switched Data. A communications paradigm in which a dedicated communication path is established between two hosts and on which all packets travel. The telephone system is an example of a circuit-switched network.

CSE   Circuit Switched Environment.

CSP   Client Server Protocol.

DECT   Digital Enhanced Cordless Telecommunications. This uses the 20MHz of spectrum between 1,880MHz and 1,900MHz and provides 120 duplex channels to provide wireless speech bearers. Enhancements to this standard now allow data services up to a data rate of 1Mbps to be supported.

DLL   Dynamic Linked Library. An executable program module in Windows that performs one or more functions at runtime. DLLs are not launched by the user; they are called for by an executable program or by other DLLs. The Windows OS contains a huge number of DLLs. With only one instance of the DLL open in memory, each DLL can be shared by all running applications.

DNA   Name of a content-authorizing tool supplied by 2.

DNS   Domain Name System. The DNS is a general-purpose distributed, replicated, data query service. The principal use is the look-up of host IP addresses based on host names. The style of host names now used in the Internet are called "domain names", because they are the style of names used to look up anything in the DNS. Some important domains are: .COM (commercial), .EDU (educational), .NET (network operations), .GOV (US government), and .MIL (US military). Most countries also have a domain name. The country domain names are based on ISO 3166. For example, .US (United States), .UK (United Kingdom), .AU (Australia). *See also* Fully Qualified Domain Name; Mail Exchange Record.

DRM   Digital Rights Management. Systems used to implement payments to which content owners are entitled when their content is involved in a network transaction.

Dual-mode handsets   Handsets that can operate over two different network standards.

EA External Application.

EAIF External Application Inter Face.

ECDS Ericsson Content Delivery System.

EDGE Enhanced Data rate for GSM Extension. An intermediate stage for introducing high-speed data to mobile networks ahead of 3G.

EJB Enterprise Java Bean. A software component in Sun's J2EE platform, which provides a pure Java environment for developing and running distributed applications. EJBs are written as software modules that contain the business logic of the application. They reside in and are executed in a runtime environment called an "EJB Container", which provides a host of common interfaces and services to the EJB, including security and transaction support. At the wire level, EJBs look like CORBA components.

EMA Electronic Message Association. A membership organization founded in 1983 devoted to promoting email, voicemail, fax, EDI and other messaging technologies. In 2001, the EMA was folded into The Open Group.

EMS Enhanced Message Service. A 3GPP standard for adding graphics and sounds to SMS messages.

ENUM Electronic Numbering.

EPOC Operating system for mobile devices. A 32-bit operating system for hand-held devices from Symbian Ltd, London, (**www.symbian.com**). Used in Psion and other hand-held computers, it supports Java applications, email, fax, infrared exchange, data synchronization with PCs and includes a suite of PIM and productivity applications.

ERP Ear Reference Point.

ESMTP A user-configurable, relay-only Mail Transfer Agent (MTA) with a sendmail-compatible syntax.

ETSI European Telecommunications Standards Institute. Standards body established to coordinate the development of telecommunications systems within Europe. These systems relate to fixed, wireless and cellular systems. Recently, systems such as GSM and DECT have been devolved to the 3GPP (Third Generation Partnership Project), a collaboration of standards bodies including ARIB (Association of Radio Industries and Businesses) from Japan.

FFS For Further Study.

FOMA   Freedom of Multimedia Access.

FQDN   Full Qualified Domain Name. The FQDN is the full name of a system, rather than just its hostname. For example, "venera" is a hostname and "venera.isi.edu" is an FQDN. *See also* hostname; Domain Name System.

FTP   File Transfer Protocol. A protocol that allows a user on one host to access and transfer files to and from another host over a network. Also, FTP is usually the name of the program the user invokes to execute the protocol. *See also* anonymous FTP.

GERAN   GSM EDGE Radio Access Network. This supports the EDGE (Enhanced Data rates for Global Evolution) modulation technique and has been specified to connect the A Gb and Iu interfaces to the CN (Core Network). The architecture allows two BSSs (Base Station Subsystems) to be connected to each other.

GGSN   Gateway GPRS Support Node. The Gateway GPRS Support Node supports the edge-routing function of the GPRS network. To external packet data networks the GGSN performs the task of an IP router. Firewall and filtering functionality, to protect the integrity of the GPRS core network, are also associated with the GGSN along with a billing function.

GIF   Graphic Interchange Format. Commonly used for compressing graphics on the Internet.

Gmb   Reference point between the BM-SC and the GGSN. It is used to give MBMS information like QoS and MBMS service area to the UMTS network.

GMSC   Gateway MSC. A Gateway Mobile Switching Centre provides an edge function within a PLMN (Public Land Mobile Network). It terminates the PSTN (Public Switched Telephone Network) signalling and traffic formats and converts them to protocols employed in mobile networks. For mobile-terminated calls, it interacts with the HLR (Home Location Register) to obtain routing information.

GPRS   General Packet Radio Service. GSM Phase II provides high-speed packet data rates up to a theoretical maximum of 170kbps. GPRS is an enhancement to cellular phone networks, as it enables them to carry packet-switched data traffic.

GSM   Global System for Mobile communications. The standard for digital cellular systems used throughout Europe and in other countries around the world, operating in several frequency bands.

GSMA   GSM Association.

GUP   Generic User Profile. The 3GPP Generic User Profile is the collection of data that is stored and managed by different entities, such as the UE, the Home Environment, the Visited Network and Value Added Service Providers, which affects the way in which an individual user experiences services. The GUP is composed of a number of User Profile Components, and an individual service may make use of a subset of the available User Profile Components.

HDML   Hand-held Device Mark-up Language. A specialized version of HTML designed to enable wireless pagers, cellphones and other hand-held devices to obtain information from Web pages. HDML was developed by Phone.com (now Openwave Systems) before the WAP specification was standardized. It is a subset of WAP with some features that were not included in WAP.

HE   Home Environment. The HE is the environment that is responsible for the overall provision of services to users.

HLR   Home Location Register. This is a database within the HPLMN (Home Public Land Mobile Network). It provides routing information for MT (Mobile Terminated) calls and SMS (Short Message Service). It is also responsible for the maintenance of user subscription information. This is distributed to the relevant VLR (Visitor Location Register) or SGSN (Serving GPRS Support Node) through the attach process and mobility management procedures, such as Location Area and Routing Area updates.

HSCSD   High Speed Circuit Switched Data. This enhances the capabilities of the current GSM network by combining time slots on the GSM radio carrier. Data rates of 14.4kbps, using one timeslot, 28.8kbps using two time slots or 57.6kbps using four time slots can be achieved. HSCSD is best suited for services such as video-conferencing.

HSS   Home Subscriber Server. This describes the many database functions that are required in next generation mobile networks. These functions will include the HLR (Home Location Register), DNS (Domain Name Servers) and security and network access databases.

HTML   Hyper Text Mark-up Language. The language used to create hypertext documents. It is a subset of SGML and includes the mechanisms to establish hyperlinks to other documents.

HTTP   Hyper Text Transfer Protocol. The protocol used by WWW to transfer HTML files. A formal standard is still under development in the IETF.

i-mode NTT DoCoMo's highly successful wireless Internet service, based on a proprietary packet bearer and using CHTML browsers in the mobile device.

ICSTIS International Committee for the Supervision of Standards of Telephone Information Services.

IETF Internet Engineering Task Force. The IETF is a large, open community of network designers, operators, vendors and researchers whose purpose is to coordinate the operation, management and evolution of the Internet and to resolve short-range and mid-range protocol and architectural issues. It is a major source of proposals for protocol standards that are submitted to the IAB for final approval. The IETF meets three times a year and extensive minutes are included in the IETF Proceedings.

IIOP Internet Inter-ORB Protocol. The CORBA message protocol used on a TCP/IP network (Internet, intranet, etc.). CORBA is the industry standard for distributed objects that allows programs (objects) to be run remotely in a network. IIOP links TCP/IP to CORBA's General Inter-ORB protocol (GIOP), which specifies how CORBA's Object Request Brokers (ORBs) communicate with each other.

IM Instant Messaging. Exchanging messages in real time between two or more people. Unlike a dial-up system, such as the telephone, instant messaging requires that both parties be logged onto their IM service at the same time. Also known as a "chatting", IM has become very popular for both business and personal use. In business, IM provides a way to contact co-workers at any time of the day, providing that they are at their computers. Because you are signalled when other IM users have logged on, you know they are back at their desks, at least for the moment.

IMPP Instant Messaging and Presence Protocol. The working group of the IETF that specializes in instant messaging.

IMPS Instant Messaging and Presence Services. An initiative to define and promote a set of universal specifications for mobile instant messaging and presence services and create a community of supporters.

IMS IP Multimedia System.

IMSI International Mobile Subscriber Identity. The International Mobile Subscriber Identity is a unique identifier allocated to each mobile subscriber in a GSM and UMTS network. It consists of a MCC (Mobile Country Code), a MNC (Mobile Network Code) and a MSIN (Mobile Station Identification Number).

IN Intelligent Network. Specialized services architecture that centralizes intelligence associated with enhanced services. Applications on an IN system provide the network with instructions on call routing and user interactions.

Interconnect   The commercial and technical arrangements under which two operators connect their networks so that customers of one have access to the customers and/or services of the other.

Interoperability   The ability of devices in different network domains to exchange traffic.

Interoperator exchange   Technical and commercial arrangements made between mobile network operators, whereby each agrees to terminate on its network messages that originate on the other's network.

IP   Internet Protocol. The Internet Protocol is the network layer for the TCP/IP Protocol Suite. It is a connectionless, best-effort packet-switching protocol.

IPSEC   IP Security. This provides a framework of open standards dealing with data confidentiality, integrity and authentication between participating hosts.

ISDN   Integrated Services Digital Network. An ISDN is an end-to-end digital network capable of simultaneous transmission of a range of services, such as voice, data and video, etc. The network is based upon 64kbps circuits.

ISMA   Internet Streaming Media Alliance.

ISO   International Organization for Standardization. A voluntary, non-treaty organization founded in 1946 which is responsible for creating international standards in many areas, including computers and communications. Its members are the national standards organizations of the 89 member countries, including ANSI for the USA.

ISP   Internet Service Provider. This is a company that provides Internet access to other companies and individuals.

ISTO   Industry Standards and Technology Organization.

ITU   International Telecommunications Union. An agency of the United Nations that coordinates the various national telecommunications standards so that people in one country can communicate with people in another country.

IVR   Interactive Voice Response. The systems that provide information in the form of recorded messages over telephone lines in response to user input in the form of spoken words or, more commonly, DTMF (Dual Tone Multiple Frequency) signalling.

J2EE Java 2 Enterprise Edition. The Java 2 Platform Enterprise Edition (J2EE) defines the standard for developing multitier enterprise applications. J2EE simplifies enterprise applications by basing them on standardized, modular components, by providing a complete set of services to those components, and by handling many details of application behavior automatically.

J2ME Java 2 Micro Edition. A highly optimized Java runtime environment, J2ME technology specifically addresses the vast consumer space that covers the range of extremely tiny commodities, such as smart cards or a pager all the way up to the set-top box, an appliance almost as powerful as a computer.

J2SE Java 2. Technology for rapidly developing and deploying mission-critical enterprise applications, J2SE provides the essential compiler, tools, runtimes and APIs for writing, deploying and running applets and applications in the Java programming language.

JDBC Java Data Base Connectivity. JDBC technology is an API that lets you access virtually any tabular data source from the Java programming language. It provides cross-DBMS connectivity to a wide range of SQL databases and now, with the new JDBC API, it also provides access to other tabular data sources, such as spreadsheets or flat files.

JPEG Joint Photographic Expert Group. An international standard format for digital compression of still photographic images.

JSP Java Server Page. An extension to the Java servlet technology from Sun that provides a simple programming vehicle for displaying dynamic content on a Web page. The JSP is an HTML page with embedded Java source code that is executed in the Web server or application server. The HTML provides the page layout that will be returned to the Web browser, and the Java provides the processing (e.g., to deliver a query to the database and fill in the blank fields with the results).

kbps A measure of transmission speed. A transmission rate of 1kbps corresponds to 1,000 bits per second.

LAN Local Area Network. A data network intended to serve an area of only a few square kilometers or less. Because the network is known to cover only a small area, optimizations can be made in the network signal protocols that permit data rates up to 100Mbps.

LCS Location services. Location services are those provided to clients, giving information. These services can be divided into: value-added services, such as route planning information; legal and lawful interception services, such as those that might be used as evidence in legal proceedings; emergency services to provide location information for organizations, such as fire and ambulance service.

LDAP   Lightweight Directory Access Protocol. This protocol provides access for management and browser applications that provide read/write interactive access to the X.500 Directory.

Legacy handsets   In the context of MMS, handsets that are not fully capable of sending and receiving messages that use the MMS standard.

Legacy support   The ability to deliver some of the content of an MMS message to a legacy handset.

LIF   Location Interoperability Forum. This defines and promotes an interoperable location services solution that is open, simple and secure.

M2P   Machine-to-Person.

M4IF   MPEG4 Industry Forum.

MAP   Mobile Application Part. A protocol that enables real-time communication between nodes in a mobile cellular network. A typical usage of the MAP protocol would be for the transfer of location information from the VLR (Visitor Location Register) to the HLR (Home Location Register).

MBMS   Multimedia Broadcast/Multicast Service.

Mbps   A measure of transmission speed. A transmission rate of 1Mbps corresponds to 1 million bits per second.

MCC   Mobile Country Code.   A three-digit identifier uniquely identifying a country (not a PLMN).

MExE   Mobile Station Application Execution Environment.

MGW   Media gateway. A gateway that supports both bearer traffic and signalling traffic.

Middleware   A term for software that sits below the application, hiding the different operating systems, databases, network systems and protocols. It provides implementation-independent programming interfaces for an application to be written to.

MIDI   Musical Instrument Digital Interface. A standard protocol for the interchange of musical information between musical instruments, synthesizers and computers. It defines the codes for a musical event, which includes the start of a note, its pitch, length, volume and musical attributes, such as vibrato.

MIDP Mobile Information Device Protocol. MIDP, combined with the Connected Limited Device Configuration (CLDC), is the Java runtime environment for today's mobile information devices (MIDs), such as phones and entry-level PDAs.

MIME Multipurpose Internet Email Extensions. An extension to Internet email that provides the ability to transfer non-textual data, such as graphics, audio and fax.

MLP Mobile Location Profile. A programming interface (API) for cellphones and pagers for the Java 2 Platform, Micro Edition (J2ME). It provides support for a graphical interface, networking and storage of persistent data for "MID Profile" applications, also known as "midlets".

MM Multimedia Message. A Multimedia Message is part of the MMSE (Multimedia Message Service Environment) and is capable of carrying such formats as text, speech, still image and video.

MMBox Multimedia Message Box. Network-based persistent storage for MMS messages.

MMA Mobile Marketing Association.

MMC Multi Media Client.

MMS handset A phone or hand-held computer that is capable of sending and receiving messages using the MMS standard.

MMS Multimedia Messaging Service. A standard defined by the 3GPP for sending messages to and from mobile phones that can contain text, graphics, photos, audio and video.

MMSC MMS Centre. The equipment installed in operator networks that is used to deliver MMS services.

MMSE Multimedia Messaging Service Environment. The Multimedia Messaging Service Environment is a collection of MMS (Multimedia Messaging Service) network elements that come under the control of a single administration. However, in the case of roaming the visited network is considered to be part of that user's MMSE.

MMSNA Multimedia Messaging Service Network Architecture.

MNC Mobile Network Code. A two-digit identifier used (like the 3-bit-long NCC) to uniquely identify a PLMN.

MNP Mobile Number Portability. The ability for a mobile subscriber to change subscription network within the same country while retaining their original MS-ISDN(s).

MO Mobile Originating. This is the term given to all communication initiated at the mobile.

Mobile e-commerce (or m-commerce) Conducting commercial transactions over mobile networks.

Mobile IP A protocol from the IETF that maintains a mapping between an .xed user IP address (the home address) and a temporary IP address (the care-of address) that can change. This is used within cellular networks to ensure no loss of communication when a user moves between cells.

MP3 Motion Picture Expert Group layer 3. A popular compression format that enables the digital distribution of music.

MPEG Motion Picture Expert Group that has developed digital compression technologies for audio storage, video storage and digital television.

MRV Mobile Rights Voucher.

MS Mobile Station. The Mobile Station is a wireless terminal enabling the user to access network services over a radio interface. The MS consists of the ME (Mobile Equipment) and either a programmed UIM (User Identity Module) or a R-UIM (Removable-User Identity Module).

MS-ISDN Mobile Station ISDN number. The phone number of a mobile phone.

MSC Mobile Switching Centre. This is a telecommunication switch or exchange within a cellular network architecture that is capable of inter-working with location databases.

MSN Micro Soft Network. Microsoft's ISP service

MSSP Mobile Services Switching Protocol.

MT Mobile Terminating. This is the term given to all communication that terminates at the mobile.

MTA Mail Transfer Agent. The store and forward part of a messaging system.

MVNO Mobile Virtual Network Operator. An organization that offers mobile services to customers, has its own mobile network code, issues its own SIM card, operates its own mobile switching centre (including home location register), but does not have its own radio frequency (spectrum) allocation.

NAPTR Naming Authority Pointer. The Naming Authority Pointer (NAPTR) record specifies a regular expression-based rewrite rule that, when applied to an existing string, produces a new domain label or Universal Resource Identifier (URI). This allows DNS to be used to look up services for a wide variety of resource names that are not in domain name syntax. For more information, *see* RFC 2915 and Internet **draft-ietf-urn-naptr-rr-04.txt**.

NCC Nokia Charging Centre.

NMIT Nokia Mobile Internet Toolkit.

OA&M Operations, Administration and Maintenance. This term refers to the processes and functions used in managing a network or element within a network.

OMA Open Mobile Alliance. The Open Mobile Alliance was formed in June 2002 by nearly 200 companies representing the world's leading mobile operators, device and network suppliers, information technology companies and content providers.

OMG Object Management Group. An international organization founded in 1989 to endorse technologies as open standards for object-oriented applications. The OMG specifies the Object Management Architecture (OMA), a definition of a standard object model for distributed environments, more commonly known as CORBA.

OS Operating System. This can be described as the software within a computer that controls basic functions. Examples of operating systems include Windows, DOS and Linux.

OSA Open Services Access. A system to provide a standardized, extendable and scalable interface that allows for inclusion of new functionality in the network in future releases with a minimum impact on the applications using the OSA interface.

OSR Open System Resources.

OSS Operations Support System. This can be described as a network management system supporting a specific management function, such as fault, performance, security, configuration, etc.

OTA Over The Air.

P2M Person-to-Machine.

P2P   Person-to-Person.

Packet switching   On a packet-switched network, the data forming a whole transmission (e.g., a voice call) is broken up into a series of packets. Each individual packet is routed and carried separately across the network, using whatever capacity is available at the time. The packets are resequenced and reassembled when they reach their destination.

PDA   Personal Digital Assistant. A hand-held device used to manage personal data, such as calendar and address book. Increasingly also used as a synonym for hand-held computer .

PDC   Packet Data Cellular.

PDP   Packet Data Protocol.

PDU   Protocol Data Unit.

PHS   Personal Handyphone System. A packet-based always-on digital cellular phone service deployed in Japan.

Picture messaging   A service whereby mobile phone users can send and receive pictures via their phones (one of the major service applications of MMS); however, some picture messaging services are based on other technologies.

PIM   Protocol Independent Multicast.

PLMN   Public Land Mobile Network. A public land mobile network (PLMN) is any wireless communications system intended for use by terrestrial subscribers in vehicles or on foot. Such a system can stand alone, but often it is interconnected with a fixed system, such as the public switched telephone network (PSTN). The most familiar example of a PLMN end-user is a person with a cell phone.

PONFA   Publish Once 'N' Fit All.

POP3   Post Office Protocol version 3.

Portal   A single interface via which a user accesses multiple services and sets of content. Portals can give access to mobile networks, or the Internet, or both.

PPC   Pocket PC.

PPG   Push Proxy Gateway.

Prepaid   The purchase by mobile phone users of credit in advance, which is then offset against usage of network services. Typically, prepaid users do not pay a monthly subscription fee and have no contract with the network operator. Also commonly known as "pay-as-you-go".

Proprietary technology   Technology in which the intellectual property is owned by a commercial organization and which can only be used by other parties under licence from the owning organization.

PS   Packet Switched. In a packet-switched network, data may be transferred by dividing it into small blocks or pieces known as packets. Each packet contains information in its header to allow it to be routed by packet switches across the network. This is a more efficient means of transferring data. An example of a packet-switched network is the Internet.

PSE   Personal Services Environment. This is an environment that assists a user in finding, adapting and using services that fulfil his needs given his personal profile, his mobility and his context.

PSS   Packet Streaming Service.

PSTN   Public Switched Telephone Network. This is a general term referring to the variety of telephone networks and services.

PTD   Personal Trusted Device. This is a small, highly portable wireless telecom terminal or PDA that is Internet-enabled and carries a digital certificate (cyber identity) of the owner.

QA   Quality Assurance.

QLA   Quark License Administrator. This is a Java applet that can reside on any Java-capable system, with no special "dongle" hardware needed. It supports back-up server recognition, the ability to run licenses in a concurrent scheme and the ability to check out licenses from the administrator for up to 27 years.

Quark   A company that supplies software licensing tools.

QoS   Quality of Service. The ability to define a level of performance in a data communications system (e.g., ATM networks specify modes of service that ensure optimum performance for traffic such as realtime voice and video).

QVGA   Quarter Video Graphics Array.

RDF   Resource Description Framework.

Receiving party pays   Charging regime whereby part or all of the charge for a phone call or message is paid by the party that receives it.

Reverse billing   A network facility whereby the person or organization that would normally pay to make or receive a message or call is not charged; instead, the person or organization that would not normally be the payer assumes the charge instead.

RFC   Request For Comments. The document series, begun in 1969, that describes the Internet suite of protocols and related experiments. Not all (in fact very few) RFCs describe Internet standards, but all Internet standards are written up as RFCs. The RFC series of documents is unusual in that the proposed protocols are forwarded by the Internet research and development community, acting on their own behalf, as opposed to the formally reviewed and standardized protocols that are promoted by organizations such as CCITT and ANSI.

Roaming   Technical and commercial arrangements between mobile operators, whereby each agrees to provide services to the other's subscribers, when they enter its network domain.

RPC   Remote Procedure Call. An easy and popular paradigm for implementing the client–server model of distributed computing. In general, a request is sent to a remote system to execute a designated procedure, using arguments supplied, and the result returned to the caller. There are many variations and subtleties in various implementations, resulting in a variety of different (incompatible) RPC protocols.

RR   Radio Resource is a protocol that controls the resources over an air interface.

RTP   Real Time Protocol. This is an Internet protocol standard that defines a way for applications to manage the real-time transmission of multimedia data. RTP is used for Internet telephony applications, it does not guarantee real-time delivery of multimedia data, since this is dependent on the actual network characteristics. RTP provides the functionality to manage the data as they arrive to best effect.

RTSP   Real Time Streaming Protocol.

SAR   Segmentation and Reassembly. A process by which a PDU (Protocol Data Unit) is split or segmented in order for it to be transported over a given medium. The segments will then be reassembled to form the original PDU.

SAT   SIM Application Toolkit. The SIM (Subscriber Identity Module) Application Toolkit function resides on GSM. It essentially enables the SIM card to drive the GSM handset, allowing a interactive exchange between a network application and the end-user.

SCF Service Capability Features. In an IN (Intelligent Network) this is the application of service logic to control functional entities in providing Intelligent Network services.

SCSs Service Capability Servers. These provide the applications with service capability features that are abstractions from underlying network functionality like Call Control, Message Transfer and Location Information.

SDK Software Development Kit. A set of software routines and utilities used to help programmers write an application. For graphical interfaces, it provides the tools and libraries for creating menus, dialog boxes, fonts and icons. It provides the means to link the application to libraries of software routines and to link it with the operating environment (OS, DBMS, protocol, etc.).

SDO Standards Development Organization.

SDP Service Discovery Protocol. The Bluetooth™ defined Service Discovery Protocol enables applications to discover which services and service characteristics are available on other Bluetooth™ devices.

Service provider A company supplying IT and/or telecommunications services over an infrastructure provided by a network operator. The company may or may not also act as the network operator.

SGSN Serving GPRS Support Node. The Serving GPRS Support Node keeps track of the location of an individual MS (Mobile Station) and performs security functions and access control. The SGSN also exists in a UMTS network, where it connects to the RNC (Radio Network Controller) over the Iu-PS interface.

SIM Subscriber Identity Module. Usually referred to as a SIM card, the SIM (Subscriber Identity Module) is the user subscription to the mobile network. The SIM contains relevant information that enables access onto the subscripted operator s network.

SIP Session Initiation Protocol. This is a signalling protocol for Internet conferencing, telephony, event notification and instant messaging. SIP was developed within the IETF (Internet Engineering Task Force) Multiparty Multimedia Session Control working group.

SLA Service Level Agreement. A contract between the provider and the user that specifies the level of service that is expected during its term. SLAs are used by vendors and customers as well as internally by IT shops and their end-users.

SMAF Synthetic music Mobile Application Format.

SMIL   Synchronized Multimedia Integration Language. A World Wide Web Consortium (W3C) standard for marking up Web content that has a sequential element (e.g., a series of still images or a video clip made up from a concatenated series of miniclips).

SMPP   Short Message Peer-to-Peer Protocol.

SMS   Short Message Service. A standard for sending and receiving short alphanumeric messages to and from mobile handsets on a cellular mobile network.

SMSC   SMS Centre. The equipment installed in operator networks that is used to deliver SMS services.

SMTP   Simple Mail Transfer Protocol. The standard used for sending email over the Internet.

SOAP   Simple Object Access Protocol. A message-based protocol based on XML for accessing services on the Web. Initiated by Microsoft, IBM and others, it employs XML syntax to send text commands across the Internet using HTTP. Similar in purpose to the DCOM and CORBA-distributed object systems, but lighter weight and less programming-intensive.

SS7   Signalling System No. 7. The ITU-T standard covering the signalling used in circuit-switched telephone networks. Also known as Common Channel Signalling 7 (CCS7).

SSL   Secured Socket Layer. The leading security protocol on the Internet. When an SSL session is started, the server sends its public key to the browser, which the browser uses to send a randomly generated secret key back to the server in order to have a secret key exchange for that session.

SSP   Service Switching Point and Server-to-Server Protocol.

Standard   A set of technologies that is supported by multiple interested parties and which is freely available for use by all parties without requiring licence payments or any other kind of remuneration to the body developing the technologies. Standards are usually developed under the aegis of non-commercial organizations, such as the ISO, the IETF or the 3GPP, but they can also come from a commercial organization or a consortium of commercial organizations.

STD   Subscriber Trunk Dialing. Long-distance dialing outside of the USA that does not require operator intervention. STD prefix codes are required and billing is based on call units, which are a fixed amount of money in the currency of that country.

SVG Scalable Vector Graphics. A vector graphics file format from the W3C that enables vector drawings to be included in XML pages on the Web. GIFs and JPEGs are the standard bitmapped formats for the Web, and SVG is expected to become the standard vector format for the Web.

TAP Transferred Account Procedure. The TAP record is used by GSM operators and data clearinghouses to exchange roaming information.

TCP Transmission Control Protocol. This is a reliable octet streaming protocol used by the majority of applications on the Internet It provides a connection-oriented, full-duplex, point-to-point service between hosts.

TDMA Time Division Multiple Access. A mobile telephone network technology.

Texting Sending and receiving text messages using mobile phones.

TFT Thin Film Transistor. The term typically refers to active matrix screens on laptop computers. Active matrix LCD provides a sharper screen display and broader viewing angle than does passive matrix.

TLS Transport Layer Security. This is a protocol such as SSL (Secure Sockets Layer) that normally provides confidentiality, authentication and integrity for stream-like connections, such as those provided by TCP (Transmission Control Protocol). It is typically used to secure HTTP (Hypertext Transport Protocol) connections and has been standardized by the IETF (Internet Engineering Task Force).

TTL Time To Live. A field in the IP header that indicates how long this packet should be allowed to survive before being discarded. It is primarily used as a hop count.

UA User Agent.

UAProf User Agent Profile. This is the set of information necessary to provide a user with a consistent, personalized service environment, irrespective of the user's location or the terminal used (within the limitations of the terminal and the serving network).

UCP Universal Computer Protocol. UCP is an ETSI standard protocol for sending SMS messages to the Network Equipment.

UDP User Datagram Protocol. An Internet Standard transport layer protocol defined in RFC 768. It is a connectionless protocol that adds a level of reliability and multiplexing to IP.

UE User Equipment. The UMTS Subscriber or UE (User Equipment) is a combination of ME (Mobile Equipment) and SIM/USIM (Subscriber Identity Module/UMTS Subscriber Identity Module).

UIQ   User Interface platform for Symbian version O8.

UMS   User Mobility Server.

UMTS   Universal Mobile Telecommunications System.

Unified messaging   A messaging environment that provides a single interface for handling all types of message (e.g., voicemail, fax and email). The available functions, such as filing and searching, can be performed on all types of message.

URI   Uniform Resource Identifer. The addressing technology that identifies every file (Web page, image, video clip, script, etc.) stored on the Internet.

URL   Universal Resource Locator. The address at which a particular set of content can be found on the WWW.

USB   Universal Serial Bus. This provides an expandable, Plug and Play serial interface that ensures a standard, low-cost connection for peripheral devices, such as keyboards, mice, joysticks, printers, scanners, storage devices, modems and video-conferencing cameras.

USSD   Unstructured Supplementary Services Data. This mechanism allows the MS (Mobile Station) user and a PLMN (Public Land Mobile Network) operator-defined application to communicate in a way that is transparent to the MS and to intermediate network entities.

UTRAN   Universal Terrestrial Radio Access Network. This is a conceptual term identifying that part of a UMTS network that consists of one or more RNC (Radio Network Controller) and one or more Node B between Iu and Uu interfaces.

VASP   Value Added Service Provider. Provides services other than basic telecommunications service for which additional charges may be incurred.

VB   Visual Basic. Popular programming language used to develop windows applications.

VHE   Vertical Home Environment. A concept for personal service environment portability across network boundaries and between terminals.

VLR   Visiting Location Register. This contains all subscriber data required for call handling and mobility management for mobile subscribers currently located in the area controlled by the VLR.

VMSC   Visited Mobile Switching Centre. This is the term given to the MSC (Mobile Switching Centre) that is serving a mobile in the VPLMN (Visited Public Land Mobile Network).

Voicemail   A system that stores voice messages, usually spoken over a telephone, in a voicemail box. These messages can be retrieved and forwarded to other voicemail boxes.

VPIM   Voice Profile for Internet Mail. A method for encoding voicemail messages as data so they can travel via the SMTP mail protocol over IP networks. VPIM uses MIME to encode messages in multiple parts, which are decoded by VPIM-compliant voicemail systems at the other end.

VPN   Virtual Private Network. This is a private network link that is carried on a public network through the use of tunnelling. It is likely that the communication will utilize encryption techniques.

VXML   Virtual Extended Mark-up Language. An extension to XML that defines voice segments and enables access to the Internet via telephones and other voice-activated devices.

W3C   World Wide Web Consortium.

WAE   Wireless Application Environment. Nominally viewed as the WAP Browser the WAP 2.0 Application Environment has evolved to embrace developing standards of the Internet browser mark-up language.

WAMS   Wireless Application Messaging Server. A wireless application messaging server enables mobile operators and third parties to offer global cross network SMS-based value added services and applications.

WAP   Wireless Application Protocol. A standard designed to allow the content of the Internet to be viewed on the screen of a mobile device, such as mobile phones, personal organizers and pagers. WAP also overcomes the processing limitation of such devices. The information and services available are stripped down to their basic text format.

WAP Forum   Founded in September 1997 by Ericsson, Motorola, Nokia and Openwave (then called Unwired Planet), the WAP Forum's aim is to provide a common platform to deliver Internet-based services specifically for use by mobile terminals.

WAPGW   WAP Gateway. A WAP (Wireless Application Protocol) Gateway accesses Web content for a mobile. In theory it is capable of converting HTML (Hyper Text Mark-up Language) pages to WML Wireless Mark-up Language) pages, but much of the content accessed from WAP Gateways has already been specially authored in WML.

WAV   Extension for a sound file.

WBXML   WAP Binary XML. This is a compact representation of XML. The binary XML content format is designed to reduce the transmission size of XML documents, allowing more effective use of XML data on narrow-band communication channels.

WCDMA   Wideband CDMA. This is a standard for 3G mobile networks.

WDP   Wireless Datagram Protocol. This is a general datagram service, offering a consistent service to the upper layer protocols and communicating transparently over one of the available underlying bearer services. It forms part of the WAP (Wireless Application Protocol) suite.

WEB   Short for World Wide Web, the Internet. A hypertext-based, distributed information system created by researchers at CERN in Switzerland. Users may create, edit or browse hypertext documents. Clients and servers are freely available.

WIM   WAP Identity Module. This is used in performing WTLS (Wireless Transport Layer Security) and application-level security functions and, especially, to store and process information needed for user identification and authentication.

Wireless email   A service providing partial or full access to a user's email inbox from a wireless-connected device.

Wireless portal   A concentration of content and applications that is available over the mobile network using a hand-held terminal.

WLAN   Wireless LAN. This is a generic term covering a multitude of technologies providing local area networking via a radio link. Examples of WLAN technologies include WiFi (Wireless Fidelity), 802.11b and 802.11a HiperLAN Bluetooth IrDA (Infrared Data Association) and DECT (Digital Enhanced Cordless Telecommunications), etc.

WMF   World Media Forum.

WML   Wireless Mark-up Language. This is a mark-up language based on XML and is intended for use in specifying content and user interface for narrow-band devices, including cellular phones and pagers.

WPAN   Wireless Personal Area Network. A wireless network that serves a single person or small workgroup. It has a limited range and is used to transfer data between a laptop or PDA and a desktop machine or server as well as to a printer.

WSDL   Web Services Description Language. A protocol for a Web service to describe its capabilities. Co-developed by Microsoft and IBM, WSDL describes the protocols and formats used by the service.

WSP WAP Session Protocol. This provides the upper level application layer of WAP (Wireless Application Protocol) with a consistent interface for two session services. The first is a connection-mode service that operates above the transaction layer protocol, and the second is a connectionless service that operates above a secure or non-secure datagram transport service.

WTLS Wireless Transport Layer Security. This layer is designed to provide privacy, data integrity and authentication between two communicating applications. WTSL forms part of the WAP (Wireless Application Protocol) suite.

WTP Wireless Transaction Protocol. This has been defined as a lightweight, transaction-oriented protocol that is suitable for implementation in thin clients and operates efficiently over wireless datagram networks. WTP forms part of the WAP (Wireless Application Protocol) suite.

WWW World Wide Web. An international, virtual, network-based information service composed of Internet host computers that provide online information.

XHTML Extensible Hyper Text Mark-up Language. The combining of HTML 4.0 and XML 1.0 into a single format for the Web. XHTML enables HTML to be eXtended (the X in XHTML) with proprietary tags.

XMF Extensible Messaging Framework. A flexible messaging architecture that allows easy modification of system behaviour through configurable scripts and rule-based decisions. Operators can adopt new services in the field without requiring new code, thereby shortening time to market.

XML Extensible Mark-up Language. The major difference between HTML (Hyper Text Mark-up Language) and XML is that HTML describes the content of a Web page and XML classifies the information. XML indicates what type of information should go where and in what format.

Standards and Specifications

3GPP TS 22.140	Services and system aspects
3GPP TS 23.140	Multimedia Messaging Service (MMS)
3GPP TS 26.140	Media formats and codecs
3GPP TS 32.235	Charging specification
WAP-205	MMS architecture
WAP-206	MMS client transactions
WAP-209	MMS encapsulation
WAP-203	WSP specification
OMA-DRM-v1	Digital Rights Management
RFC2822	Internet Message Format
RFC2387	The MIME Multipart Related Content
RFC2557	MIME Encapsulation of Aggregate Documents, such as HTLML (MHTML)
W3C SVG	Scalable Vectors Graphics
W3C SOAP	Simple Object Access Protocol
W3C XHTML MP	XHTML Mobile Profile
W3C SMIL	Synchronized Multimedia Integration Language

Multimedia Messaging Service Daniel Ralph and Paul Graham
© 2004 John Wiley & Sons, Ltd   ISBN: 0-470-86116-9

Websites

3G Americas (www.3gamericas.org)   General news site covering commercial and technical developments across the mobile industry and in relation to America.

3G News and Information (www.3g.co.uk)   General news site covering commercial and technical developments across the mobile industry.

3G News Room (www.3gnewsroom.com)   General news site covering commercial and technical reports across the mobile industry.

All Net Devices (www.allnetdevices.com)   General news site covering mobile Internet and device topics.

Ericsson (www.ericsson.com/mobilityworld)   Information for developers, including development environments and support notes.

GSM BOX (www.uk.gsmbox.com)   A European website with technical reviews of handsets and popular news stories.

MMS Links page (mms.start4all.com)   Lists of links to general MMS information sites, including phones, operators, content providers, news and software vendors.

MobileYouth (www.mobileyouth.org).

Multimedia Messaging Service   Daniel Ralph and Paul Graham
© 2004 John Wiley & Sons, Ltd   ISBN: 0-470-86116-9

Multimedia content provider (www.multimediamessaging.com)
Access to subscription-based MMS content, including backgrounds, images and storage space.

Nokia (www.forum.nokia.com)   General information, development environments and product specifications for the current range of Nokia devices.

PicturePhoning (www.textually.org/picturephoning)   Covers events and news stories within the mobile industry or uses picture messaging technology in support of innovative new applications.

Telecoms.com (www.telecoms.com)   Telecoms.com is owned and operated by the Informa Telecoms Group. It is the definitive source of high-quality information on mobile telecoms markets and technology.

WAPwednesday   (www.wapwednesday.com)   Wireless Around People is a networking organization that hosts regular wireless industry events around the globe. Since its formation in December 1999, WAPwednesday has developed a reputable and well-recognized networking platform for the wireless community.

Wireless Forum (www.w2forum.com)   General information, news, views and market reports on the wide range of mobile industry matters.

Industrial Fora, Regulatory Organizations and Other Relevant Initiatives

3GPP (www.3gpp.org) Third Generation Partnership Project in charge of developing specifications for the EDGE and WCDMA/UMTS standards. 3GPP organizational partners include ARIB (Japan), CWTS (China), ETSI (Europe), T1 (USA), TTA (Korea) and TTC (Japan).

3GPP2 (www.3gpp2.org) 3GPP2 provides globally applicable technical specifications for a Third Generation mobile system based on the evolving ANSI-41 (CDMA/TDMA) core network and the relevant radio access technologies to be transposed by standardization bodies (organizational partners) into appropriate deliverables (e.g., standards).

Fixed Line MMS Forum (www.f-mms.org) The objective of the Fixed Line MMS Forum is to introduce MMS in the fixed network in order to enhance attractiveness by adding new services (including voice data services). This will lead to fixed–mobile convergence applications while ensuring interoperability between fixed and mobile networks.

Multimedia Messaging Service Daniel Ralph and Paul Graham
© 2004 John Wiley & Sons, Ltd ISBN: 0-470-86116-9

GSM Association   (www.gsmworld.com)   The GSM Association (GSMA) is a global trade association that represents the interests of more than 550 GSM mobile operators and releases information such as statistics and latest news on MMS launches and new service initiatives.

Internet Engineering Task Force (www.ietf.org) A large, open, international community of network designers, operators, vendors, and researchers concerned with the evolution of the Internet architecture and the smooth operation of the Internet. It is open to any interested individual.

International Telecoms Union (www.itu.int/ITU-D/ict/statistics/) The ITU, headquartered in Geneva, Switzerland is an international organization within the United Nations System where governments and the private sector coordinate global telecom networks and services.

MPEG4 Forum (www.m4if.org)   MPEG4 is an ISO/IEC standard developed by MPEG (Moving Picture Expert Group), whose formal ISO/IEC designation is ISO/IEC 14496. It was finalized in October 1998 and became an international standard in the first months of 1999.

Open Mobile Alliance (www.openmobilealliance.org)   The mission of the Open Mobile Alliance is to grow the market for the entire mobile industry by removing the barriers to global user adoption and by ensuring seamless application interoperability while allowing businesses to compete through innovation and differentiation.

World Wide Web Consortium (W3C) (www.w3c.org)   An organization with over 500 members, created in 1994, that has been leading the development of protocols aimed at promoting Internet interopearability. W3C activities have included the development, standardization and promotion of languages such as XML, RDF, VoiceXML and SMIL.

References

In preparing and writing this book I have drawn on many resources, from news articles, conversations and emails to standards documentation, conference workshops and panel discussions. The following is a list of publicly available online resources, scientific articles, research reports, books and other publications that have been found useful. Given the fluidity of developments in the mobile messaging space, specific URLs have not been selected, as guarantees cannot be given that they will continue to exist.

3GPP (2002a) *3GPP TS 23.140 Release 5 Specification.*

3GPP (2002b) *3GPP TS 32.235 Release 5 Specification.*

3GPP (2002c) *3GPP TS 23.127 Release 5 Specification.*

3GPP (2002d) *3GPP TS 22.024 Release 5 Specification.*

3GPP (2002e) *3GPP TS 29.198 – 1 Open Service Access (OSA) Application Programming Interface (API); Part 1:Overview.*

3GPP (2002f) *3GPP TS 29.198 – 12 Open Service Access (OSA) Application Programming Interface (API); Part 12: Charging.*

3GPP (2002g) *3GPP TS 29.198 – 11 Open Service Access (OSA) Application Programming Interface (API); Part 11: Accounting.*

3GPP (2002h) *3GPP TS 29.846 Multimedia Broadcast/Multiservice (MBMS).*

3GPP (2002i) *3GPP TS 23.057 Mobile Station Application Execution Environment (MExE).*

3GPP (2002j) *3GPP TS 23.086 Advice of Charge Supplementary Services Stage 2.*

Multimedia Messaging Service   Daniel Ralph and Paul Graham
© 2004 John Wiley & Sons, Ltd   ISBN: 0-470-86116-9

3GPP (2002k) *3GPP TS 23.048 Security Mechanisms for the (U)Sim Applications Toolkit.*

3GPP (2002l) *3GPP TS 26.234 Transparent End-to-end Packet-switched Streaming Service (PSS); Protocols and codecs.*

3GPP (2002m) *3GPP TS 22.121 The Virtual Home Environment.*

3GPP (2002n) *3GPP TS 23.057 Mobile Execution Environment (MExE); Functional description.*

Aaltonen, S. (2002) "Web services: New opportunity for stimulating innovation in mobile content", Nokia, November (www.nokia.com/nmic2002/downloads/pdf/NMIC_Seppo_Aaltonen.pdf).

Bond, K. (2002) "Mobile content and entertainment", Analysys (research.analysys.com).

CSFB (2002) "Fatphone: A week in European mobile data", April (http://www.freever.org/press/articles_new/fatphone240402.pdf).

Davies, M. (2002) "A usability test by 3G Lab", 3G LAB (www.3glab.com/products/usabilityreport.pdf)

Delaney, J., Betti, D. and Murrell, N. (2002) *MMS and SMS: Multimedia Strategies for Mobile Messaging*, Ovum, ISBN 1-904173-08-X.

Elsen, I., Hartung, F., Horn, J., Kampmann, M. and Peters, L. (2001) "Streaming technology in 3G mobile communication systems, IEEE, September (http://www.lnt.de/~hartung/StreamingTechnologyin3GMobileCommunicationSystems.pdf)

Engdegard, S. (2002) "MMS advertising", *Jupiter Research*, December.

Ericsson (2002) "Developers guidelines MMS", October (www.ericsson.com/mobilityworld/sub/open/technologies/messaging/docs/mms_dev_guide_sony_e).

ETSI ES 201-912 (2002) *Access and Terminals (AT); Short Message Service (SMS) for PSTN/ISDN; Short Message Communication between Fixed Network Short Message Terminal Equipment and a Short Message Service Centre.*

Ferris Research (2003) "Instant messaging and presence: Market analysis 2002–2007", April (http://www.ferris.com/rep/200303/SM.html).

Frost and Sullivan (2002) "Driving MMS adoption", October (www.openwave.com/docs/products/resources/mms_oct_02.pdf).

GSM Association (2001) "M-services guidelines", May (http://www.gsmworld.com/documents/m-services/aa35.pdf)

Harris, I. (2002) "Multimedia messaging", Teleca, July (http://www.wapwednesday.co.uk/archives/London/july_02/WAPwednesday_Teleca.pdf).

Hietala, E. (2002) "FAQ: How to become an MMSC service provider", October (www.wiral.com/w_2/formmanager/downloads/files/FAQ_MMS_Service_Provider.pdf).

Hilavuo, S. (2002) "Global practices in tariffing and revenue sharing", Nokia, November (www.nokia.com/nmic2002/downloads/pdf/NMIC_Sonja_Hilavuo.pdf).

Ikola, S. H. (2002) "Mobile DRM", Nokia, June
(www.forum.nokia.com/seap/Digital_Rights_Management_Sirpa_Ikola2.pdf).

ITU (2002) "Mobile cellular, subscribers per 100 people", December
(www.itu.int/ITU-D/ict/statistics/at_glance/cellular02.pdf).

Ives, S. (2002) "Unlocking new revenue streams in mobile", 3Glab, June
(www.3glab.com/trigenixworld/unlockingrevenue.pdf).

Kantsopoulou, M., Alonisticti, N., Gazis, E. and Kaloxylos, A. (2002) "Adaptive charging accounting and billing system for the support of advanced business models of VAS provision in 3G systems", BSc thesis at National and Kapodistrian University of Athens, School of Sciences, Department of Informatics and Telecommunications.

Lannerstrom, S. (2002) "Catalyst for succesful MMS", SmartTrust, August
(www.smarttrust.com/whitepapers/pdf/mms.pdf).

Le Bodic, G. (2002) *Mobile Messaging: Technologies and Services*, John Wiley & Sons, ISBN 0-471-84876-6, pp. 197–334.

Machin, R. (2002) *MMS Billing and Accounting: Version 1*, Telecoms Solutions, April.

Newbould, R. and Collingridge, R. (2003) *BTTJ: Profiling Technology*, Kluwer, ISSN 1358-3948, pp. 13–20.

Nokia (2002a) "How to create MMS services", July
(ncsp.forum.nokia.com/downloads/nokia/documents/How_to_Create_MMS_Services_v3_21.pdf).

Nokia (2002b) "Multimedia streaming"
(www.nokia.com/downloads/aboutnokia/press/pdf/streaming_wp.pdf).

Nokia (2002c) "MMSC Application development guide", March
(http://www.forum.nokia.com/html_reader/main/1,,2090,00.html).

Northstream (2002a) "MMS pricing challenges", April
(www.northstream.se/download/mms_pricing.pdf).

Northstream (2002b) "When will video be delivered to my phone?", April
(www.northstream.se/download/mms_pricing.pdf).

Novak, L. and Svensson, M. (2001) "MMS – Building on the success of SMS", Ericsson (http://www.ericsson.com/about/publications/review/2001_03/140.shtml).

Oftel (2002) "Market information mobile update", October
(http://www.oftel.gov.uk/publications/market_info/).

OMA (2002) "WV-020 System Architecture Model Version 1.1 OMA-WV-Arch-V1_1-20021001 A"
(http://www.openmobilealliance.org/wirelessvillage/docs/WV_Architecture_v1.1.pdf).

OMA MMS-V11 "MultiMedia Messaging Service Version 1.1"
(http://www.openmobilealliance.org/documents.html).

Openwave (2002) "MM7 developers guide"
(developer.openwave.com/omdt/MMS%20SDK/_mm7_dev.pdf).

Ortiz, E. and Giguere, E. (2001) *Mobile Information Device Profile for J2ME*,
John Wiley & Sons, ISBN 0-471-03465-7.

Patel, N. (2002) "Personal content management and storage", Strategy
Analytics, October (http://www.strategyanalytics.com/)

Patel, N. (2003) "Challenges and risks in the US$700m Mobile Marketing
opportunity", Strategy Analytics, February (http://www.strategyanalytics.com/).

Penttila, N. (2003) "Personalized services for virtual home environment",
BSc thesis at Department of Computer Science and Engineering, Hel-
sinki University of Technology.

ProQuent Systems (2003) *MSSP 2800 Series Product Description*.

Ralph, D. and Searby, S. (2003) *BTTJ: Location and Personalisation*, Kluwer,
ISSN 1358-3948, pp. 13–20.

Sadeh, N. (2002) *M-commerce: Technologies, Services and Business Models*,
John Wiley & Sons, ISBN 0-471-13585-2, pp. 33–62.

Soapuser (2002) "What is SOAP" (www.soapuser.com).

Stegavik, H. (2002) "Launching third party applications with an opera-
tor", Telenor, June (http://www.nokia.com/nmic2002/downloads/pdf/NMIC_Harald_Stegavik.pdf).

Strategy Analytics (2002) "MMS market dynamics: Content to drive MMS
revenues", December (www.strategyanalytics.com/cgi-bin/greports.cgi?rid=152002120451).

Turnbull, E. and Bond, K. (2002) *Enabling Prepaid Mobile Content and Data
Services*, Analysys, June, ISBN 1-903648-36-X (research.analysys.com).

W3C SOAP 1.1 Schema (3GPP) *MMS MM7 VASP Interworking Stage 3
Specification* (http://www.3gpp.org/ftp/specs/archive/23_series/23.140/schema/)

WAP-189, OMA (2002) *Push OTA Protocol*.

WAP-205, OMA (2001) *MultiMedia Messaging Service; Architecture Overview Specification*.

WAP-209, OMA (2002) *MMS Encapculation Protocol*.

Wirzenius, L. and Marjola, K. (2002) *Kannel 1.1.6 User Guide*, Open Source
WAP and SMS Gateway.

Wisley, D. R., Eardley, P. and Burness, L. (2002) *IP for 3G: Networking
Technologies for Mobile Communications*, John Wiley & Sons, June ISBN
0-471-48697-3.

Yung-Stevens, A. (2002) "Commercialising MMS Applications with
operators", Nokia, June
(www.forum.nokia.com/seap/Commercializing_MMS_Apps_AnnaYungStevens.pdf).

Index

Multimedia Messaging Service   Daniel Ralph and Paul Graham
© 2004 John Wiley & Sons, Ltd   ISBN: 0-470-86116-9